T0345919

THE STATE OF NATURE

Science and Its Conceptual Foundations
David L. Hull, Editor

GREGG MITMAN

The State of Nature

Ecology, Community, and American Social Thought, 1900–1950

THE UNIVERSITY OF CHICAGO PRESS
Chicago & London

GREGG MITMAN is assistant professor in the history of science at the University of Oklahoma.

The University of Chicago Press, Chicago 60637
The University of Chicago Press, Ltd., London

Printed in the United States of America
00 99 98 97 96 95 94 93 92 5 4 3 2 1

ISBN (cloth): 0-226-53236-4
ISBN (paper): 0-226-53237-2

Library of Congress Cataloging-in-Publication Data

Mitman, Gregg.
The state of nature : ecology, community, and American social thought, 1900–1950 / Gregg Mitman.
p. cm.
Includes bibliographical references.
1. Ecology—Study and teaching—United States—History—20th century. 2. Science—United States—Methodology—History—20th century. 3. Science—Social aspects—United States—History—20th century. I. Title.
QH541.26.M58 1992
574.5′071′073—dc20 91-45638
CIP

∞The paper used in this publication meets the minimum requirements of the American National Standard for Information Sciences—Permanence of Paper for Printed Library Materials, ANSI Z39.48–1984.

To Pat, Jean, and Bill—
mentors and friends

Nor think, in Nature's state they blindly trod;
The state of Nature was the reign of God:
Self-love and social at her birth began,
Union the bond of all things, and of man.
Pride then was not; nor arts, that pride to aid;
Man walked with best joint tenant of the shade;
The same his table, and the same his bed;
No murder clothed him, and no murder fed.
In the same temple, the resounding wood,
All vocal beings hymned their equal God:
The shrine with gore unstained, with gold undressed,
Unbribed, unbloody, stood the blameless priest:
Heaven's attribute was universal care,
And man's prerogative to rule, but spare.
Ah! how unlike the man of times to come!
Of half that live the butcher and the tomb;
Who, foe to Nature, hears the general groan,
Murders their species, and betrays his own.
But just disease to luxury succeeds,
And every death its own avenger breeds;
The fury-passions from that blood began,
And turned on man, a fiercer savage, man.

Alexander Pope, *An Essay on Man* (1733–1734)

Contents

Illustrations

Acknowledgments

Many books begin, as this one did, as a formless manuscript that defies organization and control and has all the attributes of a disheveled, rampaging beast. The idea for this book originated in a seminar on the history of trade associations offered by J. Rogers Hollingsworth at the University of Wisconsin, and I am grateful to him for the encouragement he gave a historian of science in the midst of his more characteristic group of students. It is the late Bill Coleman, however, to whom I owe the most profound thanks. Bill helped bring the project into a more manageable and definable perspective, and his interest in and dedication to the project and commitment to me as a student at a time of adversity in his own life has been a continual source of inspiration. He encouraged me to explore the periphery of the history of biology and to leave few stones unturned. Elliott Sober, Ronald Numbers, and Chip Burkhardt complete the immediate academic support network. Their advice and criticism, and their enthusiasm for the project have been invaluable.

I owe special thanks to my colleague, John Beatty, who has the rare talent of listening attentively and patiently. I accosted John unrelentingly throughout the past year to discuss many an idea. The manuscript has also benefited from discussions with numerous other people. They include Garland Allen, Gene Cittadino, Adele Clarke, Stephen Cross, Evelyn Fox Keller, Jane Maienschein, Philip Pauly, Robert Nye, Ronald Rainger, and Peter Taylor. Sharon Kingsland and Michael Wade read the entire manuscript, and I am grateful for their comments and suggestions. In finding the title, Pam Gossin's creative talents were particularly helpful, and David Levy added an American historian's perspective. Susan Abrams's faith in the project during some rather uncertain times deserves special mention.

For their willingness to recount their own stories of Chicago ecology, I thank Willis Johnson, David Kistner, Kumar Krishna, Ernst Mayr, Thomas Park, and Gerald Scherba. Elizabeth MacMahan graciously sent me copies of her correspondence with Alfred Emerson. Sabron Newton offered much valuable information on the 57th Street Meeting of Friends, and Dorothy Allee presented me with useful material on the Allee family history. Molly Barth's recollections of her parents, Clyde and Marjorie Hill Allee, also added a touch of warmth to persons whom I only came to know through the written word.

For archival assistance, thanks go to the staffs of the American Museum of Natural History, Earlham College library, the Harvard University Archives, the Field Museum of Natural History, the Joseph Regenstein Library, Lake Forest College Library, the library of the Marine Biological Laboratory, the Rockefeller Archive Center, and the University of Illinois Archives. Kumar Krishna, Charles Myers, Sarah Penhale, Richard Popp, and Kenneth Rose were especially helpful in locating and reproducing materials. Permission has been granted by all these institutions to quote from unpublished sources. This book incorporates material from previously published articles: "From the Population to Society: The Cooperative Metaphors of W. C. Allee and A. E. Emerson," *Journal of the History of Biology* 21 (1988): 173–94; "Dominance, Leadership, and Aggression: Animal Behavior Studies during the Second World War," *Journal of the History of the Behavioral Sciences* 26 (1990): 3–16; and "Evolution as Gospel: William Patten, the Language of Democracy, and the Great War," *Isis* 81 (1990): 446–63.

Financial assistance has come in various forms. I would like to thank the Woodrow Wilson Foundation and the Rockefeller Foundation for fellowships that gave me the necessary time for writing and reflection. A National Science Foundation grant (SES-8603920) provided the funds necessary for archival research and oral history.

Friends and family have also made this enterprise possible. I am especially grateful to John Neu for the morning coffees and friendship. Tim Allen has kept me abreast of the important ideas current in ecology; his creative mind continually challenges me to look for unexplored connections in my own work. To Helen Klebesadel I owe special thanks for her many insights into feminist issues; and I admire her commitment to reach beyond the ivory tower of academic scholarship and work for political change. No words can express my gratitude to Debra Klebesadel. Her editorial skills, her patience, her interest, and her encouragement helped sustain my dedication to this project from its inception.

Nature's Many Facets

Ecology as a subject and as a discipline has come to symbolize many things. To some, it is synonomous with the environmental movement of the 1960s, of population growth control, of Aldo Leopold's "land ethic." To others, it represents a scientific disciplinc that studies interactions between organisms and their environments. Throughout the past one hundred years, ecology as a science has undergone a marked period of specialization and refinement into a wealth of subdisciplines: animal and plant ecology; physiological, behavioral, and evolutionary ecology; and population, community, and ecosystem ecology—all grouped together under a single heading. During the early part of this century, however, the disciplinary boundaries and subdivisions of ecology were less clearly defined. Physiological ecology blurred into behavioral ecology; population ecology was, at one time, a subcomponent of community ecology. Some fields, such as evolutionary ecology, were nonexistent. Nevertheless, then, as today, the motives and goals of the practitioners of ecology were diverse. Amid this multiplicity of meanings resides an attribute that captures one important nuance of ecology in its formative years. Ecology represented the borderland between the biological and social sciences through the study of interrelationships between and among individual organisms and their environment. This meaning has largely disappeared from today's disciplinary canon as ecologists incorporate sophisticated quantitative techniques and theories from the physical sciences, hoping to make ecology a "hard science." During the first half of this century, however, the social dimension of ecology played an important defining role. By studying animal and plant societies, many ecologists hoped to bring biological understanding to problems confronting human society in what seemed to be an acutely troubled time.

The belief that biology should offer guidelines for human actions and behaviors has a rich and often sordid past, a tale detailed at length by many a historian. The story usually begins with Darwin placing humankind in the kingdom of nature, a kingdom rooted in the political economy of nineteenth-century industrial capitalism where competition and the individual struggle for survival yielded a Victorian sense of natural balance and order. Despite the misgivings expressed by Thomas Henry Huxley in his famous 1893 lecture "Evolution and Ethics," the aesthetic of nature "red in tooth and claw" as the driving force behind social progress flourished in the social Darwinism of the Gilded Age, in the eugenics movement of the Progressive Era, and in the military ideology and sterilization campaigns of Nazi Germany. The particular representations of humans who best survived this struggle for existence differed slightly, but all sketches were variations on a theme. The white Anglo-Saxon male of upper middle class was most often both patron and subject of these renditions. But the canvas of biological determinism, we are told, perished in the flames of World War II with the atrocities of Nazi Germany. Recently, it has emerged like a phoenix out of the ashes, disguised in the form of sociobiology and the social Darwinist rhetoric of Reaganite politics. Such is a brief, hackneyed version of the history of biological determinism and social Darwinism. Although these historical tales serve a valuable role in exposing what Robert Young has termed the "naturalization of values," they continually draw our attention to the political use of biology by the ideological right, without addressing how biologists left of center have utilized their science and its cultural authority in similar fashion to achieve their own political ends.[1] But if we are willing to explore science *as* culture, then we must be willing to explore how a whole amalgam of noncognitive factors such as personal beliefs and cultural norms have come to shape biology across the political spectrum.

Part of the problem with the received historical view of social biology is the preeminent status that evolutionary theory and Mendelian genetics held in the disciplinary hierarchy of early twentieth-century biological science and the subsequent attention placed on these subjects by historians of biology. These fields represent the scientific bridge that spans the interstices of social Darwinism, eugenics, the modern synthesis, and sociobiology. Yet life science in the twentieth century consists of far more than heredity and evolution. Have geneticists and evolutionary biologists been the only ones to speculate on the importance of biology for human affairs? Surely not. What is missing from the story is an understanding of how biologists in the domains of embryology, ecology, and physiology—fields with more tangential connections to developments in Mendelian genetics and Darwinian evolution—contributed to discussions of the human place in nature. This book attempts to redress the lack of balance by exploring a generation of ecologists who felt they had much to say about the significance of their science for human conduct and whose work relied heavily on techniques and theories derived from physiology, developmental biology, and animal behavior. At the University of Chicago, during a thirty-year period beginning soon after the close of World War I, Warder Clyde Allee and Alfred Edwards Emerson developed a program of animal ecology centered on studying the origins, development, and organization of animal societies. As a research program, Chicago ecology played a prominent role in the professional dialogue of ecologists in America. But ecology at Chicago also functioned within the broader social discourse of American intellectuals struggling with issues concerning international peace and democratic order during the period of and between the First and Second World Wars. Indeed, this book offers an in-depth look from the perspective of Chicago ecologists at the changing social and political landscape of biology in America during the first half of this century.

Despite the rather common belief that Darwin was the sire of ecology, a close look at the origins of animal ecology at the University of Chicago suggests otherwise. Although evolution-inclined ecologists such as Joseph Grinnell and C. C. Adams paid tribute to Darwin's entangled bank, animal ecology at Chicago developed quite independently of evolutionary doctrine. The program of Chicago animal ecology was nurtured and sustained in a research environment centered on experimental embryology with a strong physiological orientation. Exploring the geographic, institutional, and cognitive terrain at Chicago in the early 1900s reveals a central feature that existed across the University of Chicago's departmental boundaries: a focus on the organism-environment relationship as an interactive process, in which the organism, through its behavior and activity, continually restructured its environment to meet new demands. This interactive model pervaded not only embryological research but also the pragmatic philosophy of such individuals as John Dewey and George Herbert Mead. It was a developmental picture that was goal directed and progressive, lending credence to one of ecology's earliest cherished principles—succession. Furthermore, because form was interpreted as a consequence of activity, this perspective supported a physiologically based animal ecology, formulated under Victor Ernest Shelford, in which behavior was seen as the primary determinant governing the distribution and development of animal life.

In his early ecological research, Allee closely followed Shelford's program for a discipline of animal ecology built on experimental science and field investigations. Graduating from the University of

Chicago in 1912, Allee returned to his alma mater as an assistant professor of zoology in 1921. Immediately after the First World War, he initiated a research program on the causes and significance of animal aggregations, on the effects group life had on individual physiology and behavior. Allee hoped this research would bridge the widening gap between the laboratory and the field, between the physiological methods of individual ecology and the descriptive natural history that characterized community studies. Yet his research also served a political agenda. During the First World War, a strong anti-German, anti-Darwinian war rhetoric surfaced within the American biological community. One version of this antiwar biology literature was derived, in part, from the social evolutionist writings of Herbert Spencer but was tailored to fit progressive democratic ideals of community and cooperation. Exposed to this tradition of antiwar biology and deeply shaken by his own experiences as a liberal pacifist during the war, Allee embarked on a research program that melded ecological science and politics into a cooperative world. Through his aggregation research, Allee believed he had found experimental evidence opposing the doctrine of war and, also, the cornerstone to a theory of sociality centered not on the family but on the association of individuals for cooperative purposes found in the most primitive forms of life.

The protagonists in this book did not see a Hobbesian nature of war against all but instead a nature of cooperation. They had seen enough war in their lifetimes to question whether the "survival of the fittest" led to social progress or to the destruction of civilization and perhaps extinction. They did not doubt that biology was, in principle, applicable to human affairs. Indeed, few American biologists during the early twentieth century did, for that would have meant rejecting the cultural authority of biology as a beacon for the conduct and future growth of American society. The real issue was whether nature's economy was as Darwin had envisioned it. For Chicago ecologists, social progress, whether at the level of animal or of human communities, was not the result of biotic struggle but was created by the association of animals in their struggle with the abiotic environment.

Animal ecology at Chicago developed in a nonhereditary environment sustained by the fields of embryology, physiology, and animal behavior. Allee's theory of social evolution mirrored that of ecological succession; both were developmental metaphors of progress that discussed evolution without any reference to hereditary mechanisms. Instead, change took place as a consequence of the historical unfolding of the social organism from the primitive to the complex. The addition of Sewall Wright and Alfred Emerson to the zoology department at Chicago in the late 1920s, however, brought an evolutionary perspective to ecology directly informed by developments in the study of heredity. In the 1930s, the population, an intellectual axis closely aligned with genetics and evolution, usurped the community ideal, heavily laden with developmental notions derived from experimental embryology and the foundation of Chicago animal ecology in its formative years. Frank R. Lillie's proposed Institute of Genetic Biology and the hiring of Alfred Emerson, Sewall Wright, and Thomas Park prompted Allee to retool his ecological ideas in the thirties. Having to contend with a hereditary mechanism for the evolution of cooperation, Allee incorporated a group selectionist model in which the population was the fundamental unit of evolutionary change. By the late 1930s, a core group of animal ecologists had emerged at Chicago, united along a common conceptual front that emphasized the study of the population as a distinct physiological and evolutionary unit. But the organicist metaphors at the heart of Chicago ecology failed to generate new avenues of ecological research in a post–World War II environment, fearful of totalitarianism and captivated by cybernetics discourse of the organism/machine. And in clinging tenaciously to the cooperative metaphors so dear to their biological and social vision while adhering to a Darwinian framework posed by the modern synthesis, Allee and Emerson could not escape the fact that Darwin's metaphor was a struggle for existence between individuals in a harsh, competitive world.

The program of animal ecology advanced at Chicago extended well beyond the bounds of internal scientific discourse. The Great Depression and the Second World War provided new opportunities for the biologist to grasp the ethical and political lessons to be learned from studying nature's economy. Emerson championed William Morton Wheeler's concept of the superorganism to legitimate his own political views about democratic society and order within the larger context of social issues confronting biologists during the war years. Certain themes integral to Chicago ecology were reflected in Emerson's social philosophy: the population as a superorganism, the importance of group selection, and the significance of cooperation as an evolutionary force. Together these themes coalesced into a vision of a cooperative world in which individual freedom and social control functioned to ensure a stable social order. In this regard, cooperation was important as an integrating mechanism that reduced social conflict, helping to maintain social equilibrium. Cooperation was, however, not the only biological mechanism that led to increased homeostasis. Patterns of dominance and subordination were also significant in uniting the fragmented sectors of the natural and political economy into efficient and organized wholes.

Throughout the 1940s, however, the use of the superorganism

concept by Emerson and his colleague Ralph Gerard in defense of a naturalistic ethics came under increasing scrutiny among biologists outside Chicago. The paleontologist George Gaylord Simpson attacked what he called the "aggregation ethics" of the Chicago school, fearing that this brand of "totalitarianist biology" was threatening to undermine the importance of the individual at the root of democracy. Higher levels of integration, in this case, the population or community, could only be achieved by suppression of the more subordinate levels, namely, the individual. The group, Simpson argued, had no distinctive properties of its own; it was a mere amalgamation of individuals. Simpson's critique signaled a prescient warning about the fate of Chicago ecology in the postwar years. In the Cold War period, as the specter of totalitarianism in the Soviet Union was increasingly seen as a threat to democracy, the meanings and appeal of cooperation and community took on more ominous tones. As historian Richard Pells has suggested, where the "search for community" and the "need for collective action" had "captured the imagination of the Left in the 1930s," the "search for identity" and resistance to "the pressures of conformity" became dominant motifs of American liberalism in the 1950s.[2] Community was transformed into conformity, something to be resisted at all costs. The economy of nature was shifting with the changing direction of American liberalism in the 1950s: nature was competitive and the individual was the primary locus of evolutionary change.

While Emerson and Gerard utilized organicism to bolster the politics of managerial capitalism, Allee continued to build a natural edifice in the 1940s around the politics of peace in which cooperation was itself the principle biological and social goal. In the mid-1930s, Allee began work on dominance-subordinance hierarchies in the vertebrates to unravel the physiological and psychological factors at work in well-established social groups. Here, however, Allee came face-to-face with more mainstream interpretations of sociality structured around hierarchies, sex, and the family that came into direct conflict with his own theory on the origins of sociality. He did not, as Emerson did, accept dominance and subordination as a viable model for human social life, although Allee did accommodate the presence of dominance and competitive interactions within his cooperative world by resorting to group selection as the causal explanation of dominance hierarchies. Allee looked instead to the more primitive groupings of animal life as the biological pattern for human society, to the animal aggregations formed in the struggle between the organism and its abiotic environment, where division of labor was notably absent. He thus challenged the division of labor/hierarchy concept at the root of biological/social theory and early twentieth-century capitalist society. Surveying the biological literature on dominance, leadership, and aggression during the Second World War, one becomes aware of the distinctive underlying biological and political foundations of Allee's animal behavior research while remaining cognizant of the common cultural context in which Allee's experiments on vertebrate social organization took place.

Allee and Emerson, like the recent proponents of sociobiology, were trying to biologize human sociality. For them, ecology formed the basis of a scientific naturalism that functioned as both a legitimating force and a prescription for their own ethical and political views about the nature and government of human society. Nature was for them normative. It was a place to seek solace from the ethical and social problems facing a generation of American biologists whose lifetimes spanned two world wars and a devastating economic depression. It was also a source of hope and inspiration with respect to what the future might be: a place in which eternal peace and harmony were possible. And the role of the ecologist in this natural order was that of healer. Working within the constraints of nature, the ecologist could apply the knowledge derived from studying nature's economy to restore the health and survival of the diseased social organism known as human society. The task of the biologist was to discover nature's moral prescriptions and thereby serve as savior of society.

In the postwar years, this image of ecologist as social healer largely disappeared, replaced by that of the ecologist as environmental engineer. A new generation of researchers, armed with techniques such as radioactive isotopes and simulation modeling explored the biotic-abiotic interface of the ecosystem, seeking predictive power in order to assess and control the impact of industrialized societies on the environment, an environment that was showing increasing signs of stress. Ecosystem ecology, attached to the purse strings of the Atomic Energy Commission, quickly rose to prominence after the war, primarily through the efforts of Eugene Odum at the University of Georgia and Stanley Auerbach and others at the Oak Ridge National Laboratory. And the cybernetic focus of ecosystem research, with its emphasis on efficiency of information transfer between system components, had a special appeal to bureaucrats in Washington who began to sense the rising interest in environmental issues during the 1960s among their constituents. Nature's economy had to be brought under more efficient management and control.

Allee and Emerson were generally not interested in ecology as environmentalism; they rarely spoke of healing the land or of enacting conservation measures. Their interest was primarily in what the study of animal societies could tell one about *humans*. They did not want to heal nature as much as be healed by it. In their quest for a scientific humanism, nature offered meaning and reassurance. Although this ideal may have largely disappeared from ecology, it continues to play an important role in the study of animal behavior, and it is here that Allee and Emerson have had the most lasting influence.

Sociobiology is the latest version of this biological humanism; Howard Kaye aptly titles it "the natural theology of E. O. Wilson."[3] Yet while the style of argument and the methodological approach are nearly identical in the work of Allee, Emerson, and contemporary sociobiology, there is a fundamental difference in underlying assumptions about the economy of nature. Sociobiologists view the animal world in largely competitive terms; individuals are by nature selfish and strive to maximize their own reproductive success. Hence, a leading problem for sociobiology is to explain how altruistic behavior has evolved, given that individual behavior is ultimately selfish.[4] As Robert L. Trivers has argued, the task for such evolutionary models is to "take the altruism out of altruism," to "show how certain classes of behavior conveniently denoted as 'altruistic' (or 'reciprocally altruistic') can be selected for."[5] Allee and Emerson saw the animal world in cooperative terms. Altruistic behavior was not something that needed to be explained away; instead, cooperation was a fundamental evolutionary force. Thus the task for these ecologists was how to account for competition within a largely cooperative world. And just as sociobiologists reduce altruism to a form of selfish behavior, so Allee and Emerson interpreted competition as a cooperative force. Nature's economy has undergone a marked transformation in the post–World War II period, as notions of cooperation and community became limited to the confines of the nuclear family and competition came to be seen as the preserving force of a pluralistic democracy. But the appeal to nature for ethical and political wisdom remains the same.

One must ask to what extent morality is discovered in nature or is merely a reflection of the biologist's own political, social, and religious experience. Can we understand the economy of nature in a literal sense, or is it a metaphor, symptomatic and constitutive of social, conceptual, and political forces operating within a given time period? For the biologist committed to realism, nature's economy must be understood literally. To suggest otherwise would be to undermine the notion of scientist as independent, objective observer. The historian of science interested in cultural critique, however, has more license. In their analysis of gender metaphors and analogies in science, feminist scholars have provided an important theoretical framework for understanding the ways in which scientific language reifies cultural biases, transporting scientific authority into the political realm. Treating nature's economy as metaphor thus allows one to delineate the cultural context from which animal ecology at Chicago sprang and within which it functioned.[6] But as the metaphor is assimilated into scientific discourse, it is also transformed. Shorn of its history and subjectivity and buttressed by the authority and power that science conveys, the metaphor is no longer recognizable as such. It becomes dead—literally true. Therein lies the danger. We must be ever conscious that dead metaphors have etymologies; they have a history that can be critiqued and evaluated.[7] In uncovering the metaphors of nature's economy, we discover a language that has been derived first and foremost from political discourse. And because the meanings of nature's economy reside in the political arena, no easy resolution, no objective criterion can resolve such debates. We must stop looking to nature for reassurances about humanity, for we will inevitably see—as in the proverbial mirror in the story of Snow White and the seven dwarfs—a reflection of what we want to see. To shatter the mirror is to recognize that the natural, as Marx wrote, as something apart from human motivations and experiences, ceased to exist once humans entered the stage.

2

Environmental Interactions

In the summer of 1908, a young man, twenty-three years of age, stepped off the train at Chicago's Hyde Park station, having come to Illinois to begin graduate school in the Department of Zoology at the University of Chicago. Warder Clyde Allee, a youth with a stalwart frame developed from years of farm labor outside Bloomingdale, Indiana, and from afternoons and weekends spent playing football at Earlham College, walked through the gate archway into Hull Court where the University of Chicago's zoology building was housed (fig. 2.1). The cloistered walkways and gothic architecture that had characterized the university since its founding in 1892 stood in marked contrast to the expansive fields and woodlands that composed the landscape of Allee's western Indiana home. Yet Chicago's far-reaching intellectual atmosphere promised to open a frontier more expansive than the sheltered intellectual environment that contrasted with the open fields of Allee's childhood. Allee, caught in a whirlwind of new ideas and carried by the force of a burgeoning new career, would find the roots of his strong religious upbringing challenged; at the same time, his research and participation in the growing field of animal ecology would help determine the shape and form of ecological research at Chicago for forty years to come.[1]

When Allee enrolled in the summer quarter at Chicago in 1908, he had completed a year of teaching at a high school in Hammond, Indiana. For the next two years, he studied part time in the zoology department while continuing to teach high school, taking the majority of his graduate courses during the summer months. In his many sojourns on the train from Hammond to Hyde Park, Allee traveled through and near parts of Chicago that offered a glimpse into its geologic past. Throughout this region were traces of what Chicago looked like when, during the late Pleistocene epoch, the last Wisconsin ice sheet retreated north, leaving behind a body of water known as Lake Chicago; it was a lake that covered the entire Chicago plain, extending approximately fifteen miles west of the present Lake Michigan shoreline. The great ice

Figure 2.1. The University of Chicago zoology building from the southwest. The gate archway into Hull Court stands on the left-hand side. Courtesy of the University of Chicago Archives.

sheet left its mark on the topography of Chicago, altering soil conditions, drainage patterns, and exposure. These changes in turn affected the flora and fauna of the region. Tramping across the ancient beaches near Tolleston, or roaming the dunes along the south side of Lake Michigan, one could recapitulate the developmental history of Chicago and its environs. Here was a laboratory of nature where an ongoing experiment, begun thousands of years ago, awaited investigation.[2]

The Chicago landscape, as described by Rollin D. Salisbury and William C. Alden in 1899, was not a static structure frozen in time but was instead an ever-changing series of forms.[3] Dynamic processes, such as erosion and deposition, shaped the topography, and the contour of the land itself influenced the future course and direction of activity. The annual spring rains, for example, would fall on unstable soil to produce an embryonic ravine that would grow to become the main drainage conduit for the region, foreshadowing the future flood plain of a major river system. This continual interplay of process and structure resulted in development and growth, an interplay placed within a Spencerian backdrop of progress from the homogeneous to the heterogeneous, from the primitive to the complex. The belief that within this flux of activity a "general, orderly, directed development" could be discerned was a common thread that pulled together scientists in the botany, zoology, and geography departments at the University of Chicago during the 1890s and early 1900s. Philip Pauly has characterized this belief as "progressive evolutionism" and suggests that a hidden clericalism pervaded Chicago biology at the turn of the century.[4] Nature was the biologist's temple. By discovering the grand unifying order of nature, the biologist could provide a sense of purpose and meaning for humankind. Human society marked the latest stages of development, but the overall direction of this evolutionary path was not completely assured. Organisms in the past became extinct or degenerated into a parasitic existence. Only through an understanding of nature's laws and the presence of the biologist's guiding hand could one be certain that the future direction of human society was proceeding along progressive and not regressive lines. This outlook was by no means confined to Chicago; the gospel of biology was a theme present to a greater or lesser extent in writings of many noted American biologists during this period: Edwin Grant Conklin, David Starr Jordan, Vernon Kellogg, William Patten, and William Ritter, to name a few. Chicago, however, provides an important institutional focus for exploring how these notions of development, growth, and activity were played out in specific departments in different ways.

To the University of Chicago geologist Thomas C. Chamberlin or his colleague in geography, Salisbury, development proceeded first and foremost through the physical landscape. Organisms were mere pawns of the abiotic environment, and their future depended solely on the direction and course of the earth's history. Chamberlin and Salisbury accommodated a progressive view of evolution by stressing that the development of the earth was itself a story of growth. But their environmental determinism had little appeal to biologists in the zoology department at Chicago. Were organisms themselves not active players in the development of life on earth? Did they not, through their activities, condition the environment so that the future course of their evolution and development was altered? To what extent was the behavior of the organism itself a causal factor in development? Was organic structure predetermined in the hereditary constitution of the individual, or did the interaction between the developing protoplasm and environment influence the overall pattern produced? Did an organism's behavior constitute a process of adjustment to surrounding environmental conditions, or were animals mere instinctive machines, hardwired to give a precise response to a given stimulus? Such questions, which revolved around key ideas of activity, growth, and development, were at the center of turn-of-the-century biological research at the University of Chicago. And the answers offered to these questions provided an important intellectual direction to the form and content of animal ecology at Chicago during Allee's graduate days. By the time Allee entered graduate school in 1908, a program of ecology at Chicago, centered on behavior as a primary determinant governing the distribution and development of animal life, had already emerged.

A Landscape of Time

Development can be viewed from the standpoint of geological epochs, hundreds of thousands of years, or from a time frame of minutes, hours, and days, in which the growth stages of an individual organism take place. The Chicago landscape offered a developmental picture stretched over geologic time; and Chamberlin and Salisbury were two individuals who did the most to interpret the *physical* expanse of the Chicago region within a dynamic framework. Both products of the Midwest, Salisbury and Chamberlin formed a working partnership that began at Beloit College in the 1870s when Salisbury enrolled in one of Chamberlin's geology courses. When Chamberlin left Beloit in 1884, Salisbury took his former post, only to join forces with him again in 1891 as professor of geology at the University of Wisconsin where Chamberlin served as the university's president. Chamberlin's presidency was, however, short-lived. Attracted by the appeal of a university committed to graduate education and research and weary of administrative work, Chamberlin left Wisconsin in 1892 to head the Department of Geology at the newly established University of Chicago. Salisbury accompanied Chamberlin once more, this time as professor of geographic geology. During their Chicago years the two built coordinated departments, bridging the fields of geology and geography, and coauthored numerous textbooks for students in their respective departments. In 1899, Salisbury became dean of the Ogden Graduate School of Science, and went on to create in 1903 the first separate department of geography in the country to offer a doctoral degree.[5]

The separation of geography from the Department of Geology at Chicago was symptomatic of the growing professionalization of geography as an independent field during the 1890s and early 1900s. In search of a unifying concept that would demarcate the science of geography from its corollary disciplines such as geology, botany, and zoology, turn-of-the-century geographers rallied around what William D. Pattison has termed the "environmental influence idea."[6] The essence of this concept was to understand the ways in which "earth features and earth resources" influenced "the distribution, character and activities of life." Hence, "modern geography," wrote Salisbury, "is concerned especially with the effects of physical features, such as land forms, water, and climate, on living things." Or, as Walter S. Tower, a colleague of Salisbury's in the geography department at Chicago, remarked, the "essential principle of geography is relationship between physical environment and the environed organism."[7] For most American geographers in the pre–World War I era, however, the relationship between organism and environment was not mutually interactive; rather, the physical environment was the driving force behind organic change. This stringent adherence to environmental determinism placed physical geography, or physiography, at the central core of geography; and from this core, subdivisions—phytogeography, zoogeography, economic, commercial, historical, and political geography—were built.

Chamberlin and Salisbury devoted their careers to the study of physiography, which attempts to explain present-day landforms on the basis of past geologic processes such as glaciation, erosion, and deposition. Although Chamberlin and Salisbury embraced environmental determinism, they avoided the rather pessimistic picture that followed if humans were subject to the whims of the earth's forces.[8] Around 1900, Chamberlin and the University of Chicago astronomer, Forest R. Moulton, advanced the planetesimal theory, an optimistic account of the earth's origin and development that avoided the inevitable heat death of the universe portrayed in Laplace's nebular hypothesis. According to Chamberlin and Moulton, the history of the solar system was not one of degeneration from an original hot gaseous nebula that cooled but was instead a story of growth. The point of origin for their theory began with a spiral nebula composed of tiny particles, known as planetesimals. As the orbits of the planetesimals crossed paths, the planetismals grew in size and gradually joined with one another to form the present-day planets and their satellites. As the mass of the earth continued to grow, compression from its own gravity resulted in increased pressure and heat; a period of volcanic activity ensued. According to Chamberlin and Moulton's theory, the earth later entered a gradational phase of "geologic rhythm," where periods of diastrophism or deformation elevated the land, forming mountain ranges. These disfiguring periods were followed by intervals of gradual erosion, in which the once majestic areas were reduced to a peneplain, and the rocky substrates of mountainous terrain were deposited into the sea. As sediments accumulated, the sea levels rose and the land receded, to be followed by another period of diastrophic changes.[9]

Progress results from this continual pulse of activity. The physical landscape creates barriers that isolate living forms from one another so that "provincial" species, adapted to local conditions, develop. Once these barriers shift, however, or climatic conditions change, organisms once separated from one another are thrown together in the "cosmopolitan" melting pot. Competition increases, and the more mentally or socially developed organisms survive; hence, a "higher stage of adjustment" is reached. For example, Chamberlin and Salisbury argued that in the Pliocene period an "isthmian thoroughfare" between North and South America appeared, allowing the intermigration of continental fauna. The provincial South American mammalian fauna that had evolved "in relative freedom from the severe stimulus of effective competition, powerful carnivores, and shifting geographic relations" now found itself in competition with North American forms. "The carnivores of both continents throve," the pair argued, "and put a severe tax on the herbivores, forcing further progress in the line of alertness, sagacity, speed and defence, and gaining similar qualities themselves."[10] Similarly, in the Pleistocene period, when glaciers covered much of North America, increased migration of fauna to the south resulted in "an unwonted commingling of plants and animals, for every aggressive form pushed forward in the van of the advancing zone, and hence came into new organic environment, while every laggard fell behind, and was overtaken by less reluctant immigrants."[11] Salisbury and Chamberlin were most interested in understanding how changes in the physical landscape affected the development and progress of organic life. In their view, the physical environment, either through geographic isolation or climatic fluctuations, precipitated a change in conditions to which organisms had to respond or face extinction. And because the earth had originated through a process of growth, Chamberlin and Salisbury looked optimistically toward its future where "an immeasurably higher evolution than that now reached, with attainments beyond present comprehension, is a reasonable hope."[12]

Despite Salisbury's rhetoric, exploration of the impact of the physical environment on the *organic realm* was, at Chicago, left largely to departments that existed independent of geography. In fact, when Chicago's Department of Geography was first created, it cross-listed a number of courses offered by other departments. Of the proposed twenty-five courses in the new geography department, twelve were actually taught by other departments: five courses in geology, five in botany, one in zoology, and one in political economy.[13] The large number of offerings in geology is not surprising since Salisbury already held a geology appointment. Why did botany play such a large role in

Salisbury's program? The primary reason centers on one person—Henry Chandler Cowles. Cowles entered graduate school in the geology department at Chicago in 1895 and utilized the tools of physiography to fulfill Salisbury's vision of understanding the impact of the physical environment on organic life.[14]

Cowles received his B.A. at Oberlin College in 1893, having studied geology and botany under A. A. Wright. Although the major part of Cowles's first year of graduate work at Chicago was in geology, he continued to pursue his interests in botany by attending the lectures of John M. Coulter. Coulter was at the time president of Lake Forest College and commuted to the South Side three times a week to give botany lectures at the university. In 1895, a gift of $1 million from Helen Culver for the biological sciences precipitated the creation of an official Department of Botany at Chicago in 1896, at which time Coulter resigned his post as president of Lake Forest and accepted an offer from William Rainey Harper to head the newly created department. One result of the Culver gift, the Hull Biological Laboratories, a group of four buildings that separately housed the departments of zoology, anatomy, botany, and physiology, was completed in 1897. In the interim, between 1894 and 1896, students interested in botany gathered in the confined quarters of the Walker Museum where Coulter instructed them in subjects ranging from plant morphology and taxonomy to ecological relations.[15]

Coulter, trained in plant taxonomy under Asa Gray, was founder of the *Botanical Gazette*, an important journal for keeping botanists abreast of recent developments in their discipline.[16] During the 1880s and 1890s, significant changes were taking place in the professionalization of botanical science. Following the lead of Europeans, American botanists abandoned their preoccupation with description and classification, the methodological hallmarks of taxonomy, and shifted their interests to problems in experimental botany and plant physiology. Botanical research at the University of Chicago embodied the experimental precepts of the "new botany"; Coulter redirected his own research away from traditional taxonomy toward experimental morphology. The "new botany," with its focus on plant physiology and functional response, also fostered interest in the nascent science of ecology, a connection explored in depth by Eugene Cittadino.[17] Although the term oecology was introduced by Ernst Haeckel in 1866 to denote the "whole science of the relations of the organism to the environment," use of the word and systematic investigation of the subject remained sporadic until the 1890s. For many botanists, ecology, with its focus on the responses of plants to environmental influences, represented a field approach to laboratory physiology. When the Madison Botanical Congress adopted the word ecology in 1893, it was at the suggestion of the Committee on Terminology of Physiology.

Coulter was himself interested in this new science of ecology, and in 1900 he published an elementary text organized around the theme of plant relations for use in John Dewey's Laboratory School.[18] Coulter's most significant contribution to the field, however, stemmed both from a review he wrote in 1896 of the German edition of Eugenius Warming's *Plantesamfund* and from his having introduced Cowles and others in his botany classes to Warming's ideas. Warming's study of plant distributions signified a major turn from classic descriptive plant geography; his point of departure was not taxonomy but physiology. For Warming, the primary question to ask with respect to plant distribution was why different species within a given locality exhibited similarities in function and form. The chief factor responsible for an assemblage of similar but unrelated plant forms was, according to Warming, the water content of the soil. He thus classified the plant communities of his native Denmark into hydrophytes, xerophytes, and mesophytes. The xerophytic communities on the Danish sand dunes, for example, are characterized by fleshy annuals adapted to the dry, exposed, desertlike conditions along the coast. But these communities are not static entities. Changes in soil conditions, in light and shade, and in human activity all altered the environmental conditions, disrupting the ecological advantage of a species, so that a development or "succession" of plant communities could be traced over time.[19]

In Warming's work, Cowles found a potent prescription for unraveling the physiographic conditions influencing the distribution of plant life. Awarded a Ph.D. in botany and geology in 1898, Cowles combined geological principles of development with botanical science to arrive at a physiographic model of vegetational change. His dissertation, "The Ecological Relations of the Vegetation of the Sand Dunes of Lake Michigan," published in 1899, was one of the first seminal studies on plant succession in America and initiated a generation of later research.[20]

Just as Salisbury looked to the physical landscape and saw a developmental history through time, Cowles pierced the vegetational form of a given region to unveil its historical past. Ecology, Cowles wrote, is a "study in dynamics," and the task of the ecologist is to discover the laws governing developmental change. The Indiana dunes were ideal sites for this task because of their instability; they signified the early embryonic conditions of the vegetational landscape unhampered by the activities of time. In his dissertation, Cowles described the plant assemblages in order of development, beginning with the fleshy annuals that dominate the primitive beach formation close to shore and ending with the basswood, evergreen, and oak forests, which mark the established dunes farther inland. A directionality was apparent in this development: the xerophytic windblown conditions of the beach gradually gave way to the more moderate mesophytic conditions that characterized the inland plant communities. Oak vegetation grew atop a ridge that was once an ancient beach of Lake Chicago. Past and present were intertwined. The mesophytic white oak–red oak–hickory forest prefigured what the established dunes closer to the lake would become. And the vegetation of the wandering dunes near the lakeshore was indicative of the flora that marked beaches during the last ice age. As one moved away from the lake, a horizontal succession of plant forms approximated the vertical succession of vegetation over time.[21]

Although Cowles's paper, "Ecological Relations," is heralded as a classic in plant community ecology, it reveals little of the extent to which his ideas were shaped by Salisbury and Chamberlin's physiographic program. In the second bulletin of the Geographical Society of Chicago, which immediately followed Salisbury and Alden's 1899 study of Chicago geography, Cowles outlined the scope of what he called physiographic ecology. Plant distributions, Cowles argued, could be studied on either a local or regional basis. Ecological plant geography focused on the regional level, where climatic factors were the significant determinants of plant distribution. The unit of analysis for such an approach was the formation, an idea first advanced by the German phytogeographer August Griesbach and applied to the American grassland by Frederic Clements and Roscoe Pound in *The Phytogeography of Nebraska* published in 1898. Cowles chose instead local units of vegetation, where edaphic factors such as soil, slope, and light, "factors that are largely due to the physiographic nature of the district," were the causal factors governing vegetational change.[22]

Cowles divided the Chicago region into three topographic and soil areas based on Salisbury and Alden's study of the region: these included the morainic clays deposited by the last glacier; the Chicago plain, marking the former location of Lake Chicago; and the dunes and beaches of the Lake Michigan shoreline. The Chicago district had "three vegetation types connected with these three soil and topographic types: the mesophytic upland forests of the morainic clays, the hydrophytic lakes and swamps or mesophytic prairies of the Chicago plain, and the xerophytic forests of the dunes and beaches." Hence, the physical landscape determined the local vegetation through such factors as exposure, drainage, and humus content of the soil. The processes involved in plant succession, Cowles argued, were to a great extent the physiographic processes involved in erosion and deposition. Development toward mesophytic conditions could be translated into the geological language of Salisbury and Chamberlin in which denudation and deposition transform young topography into the base level or peneplain. Cowles later admitted that vegetative cycles were not inextricably linked to cycles of erosion, thereby giving some importance to biotic factors such as competition. Nevertheless, the physical environment was, by and large, the driving force behind organic change.[23]

Cowles published little after 1901, and his influence on the development of ecology at Chicago came primarily through his role as teacher; he developed a curriculum in plant ecology and took students on numerous field excursions across the country. Cowles remained at Chicago for his entire professional career, rising through the ranks from instructor in 1902 to full professor in 1915. By 1900, plant ecology courses he offered included ecological anatomy, field botany, elementary ecology, geographic botany, physiographic ecology, and a research seminar on plant ecology. All of these courses were listed in conjunction with the geography department established by Salisbury in 1903.

Cowles's vision of plant ecology fit well within Salisbury's geographic program because it emphasized the impact of the physical environment on organic life. Indeed, in Cowles's course description of field botany, he stressed that "as much attention as possible will be paid to the more purely geographical and geological features in their relation to the vegetation."[24] Missing in this outlook, however, was the role of the organism itself in its interaction with the environment. Cowles never examined the physiological response of plants to their environment, despite the hopes and rhetoric of many botanists that ecology would represent a field approach to laboratory physiology. He was not alone, however. As Joel Hagen notes, even the research of the influential plant ecologist Frederic Clements "owed more to descriptive natural history . . . than to the experimental methods of the physiology laboratory."[25] Not surprisingly, the botanist Norman Taylor, surveying the field of plant ecology in 1912, remarked that "most of the men engaged in ecological work have laid more emphasis upon the physiographic side of the subject than upon the individual response of plant organs."[26]

The physiographic emphasis of plant ecology at Chicago owed much to Salisbury and Chamberlin's influence on Cowles. Inherent in their program was a belief in development, progress, and growth that was intimately connected to the conditions of the physical environment. Chamberlin's origin of the earth hypothesis avoided the pessimistic consequences of environmental determinism by stressing that the development of the earth was itself a story of growth. Cowles added an organic dimension to the story, but still the progress and development of plant life was almost wholly dependent on inorganic change. What of the organism itself as an influence on development and progress? William Coleman has argued that a number of individuals reintroduced the psychic element of Lamarckian evolution in the late nineteenth century to combat the fatalistic picture of environmental determinism.[27] Sensate organisms, unlike plants, were able to respond to their environment, adjusting their structure and behavior to meet new conditions. Hence, organisms themselves could be active determinants of their environment. The development of the psychic element in organic life profoundly altered the relationship between organism and environment, so that animals and humans were no longer slaves to their physical surroundings. Chamberlin and Salisbury themselves looked to a psychozoic era in the future—an "eon of intellectual and spiritual development comparable in magnitude to the prolonged physical and biotic evolutions"—but the behavioral response of the organism as modifier of its environment was an aspect that they never really stressed.[28] Even J. Paul Goode, an economic geographer hired by Salisbury to help establish the geography department at Chicago, argued that, while the social environment takes on a greater role in human progress, one could never reduce the influence of the physical environment to "zero." "Be we never so wise and ingenious," Goode wrote, "we shall always be directed, and the course of our evolution will be conditioned by its [i.e., the physical environment's] elements."[29] The significance of the *organism* in changing its environment and in directing the course of its development was, however, elucidated in the zoology department at Chicago and in the program of animal ecology that came to fruition. In this setting, the behavior of the organism was a central element of progress and developmental change.

The Physiology of Development

In November 1901, with plans for a future geography department in mind, Salisbury as dean of the Ogden Graduate School of Science wrote to Charles B. Davenport, a professor in the zoology department, for his help in coordinating a course on the geographic distribution of animals. Davenport replied "that the Department of Zoology has been for some time considering giving a course in Zoogeography. I am even now preparing the lectures which are to be substituted for that on 'regeneration' this Spring. In this way Zoology will be able to aid the Geological Department as Botany does." Davenport was, however, less polite to the university's president William Rainey Harper regarding Salisbury's request. Fearing perhaps that geography under Salisbury's direction would overstep its bounds, Davenport sarcastically remarked that "since the animals . . . we are acquainted with are or have been inhabitants of the earth it is conceivable that Zoology might be considered a department of Geography. But if Zoology has any right to exist as an independent science it has a right to include in its province the subjects of the laws of occurrence of animals in space and in time–the geographical distribution of animals and their evolutionary history as recorded by their remains."[30] As this brief exchange indicates, animal ecology in the zoology department at Chicago did not bear the same relation to Salisbury's geographic program as plant ecology did under Cowles. Indeed, animal ecology at Chicago, although strongly influenced by Cowles's work, came to embrace a slightly different developmental metaphor, one centered on the interaction between protoplasm and environment in the ontogeny of the individual organism. This vision of development had its roots not in physiography but in an ongoing tradition of embryological research.

Charles Otis Whitman directed the zoology department and the biological sciences division at Chicago from its establishment in 1892 until his death in 1910. Whitman received his doctorate under Rudolph Leuckart at Leipzig in 1878 and then proceeded to move, over the next ten years, from the Imperial University in Tokyo to the Museum of Comparative Zoology at Harvard to the Allis Lake Laboratory near Milwaukee, Wisconsin, before securing an appointment as chair of zoology at Clark University in 1889.[31] Within a few years, however, financial difficulties and strained relations between Clark's president, G. Stanley Hall, and its benefactor, Jonas Gilman Clark, resulted in a number of faculty resignations. The growing dissent among faculty members at Clark proved fortuitous for the University of Chicago's first president, William Rainey Harper. Actively seeking senior faculty for his own institution, Harper literally raided Clark to acquire three new department heads, including Whitman in biology, Albert Michelson in physics, and John Nef in chemistry. In addition, an entourage of junior faculty and students accompanied these three to the Midwest.[32]

Whitman had formulated his own plans for biological instruction while still director of the Allis Lake Laboratory in 1887. In an action that reflected his own graduate education within the late nineteenth-century German university system, Whitman called for a graduate curriculum dedicated to research and specialization. At Chicago he hoped to divide biology into separate departments of zoology, botany, paleontology, physiology, and anatomy to encourage specialized research; yet he wanted to ensure coordination among the various departments. Harper saw things differently, and for the first year all the biological disciplines were under a single department. The arrangement proved temporary, and by 1893 the biological division had divided into separate departments of zoology, paleontology, physiology, anatomy, and neurology.[33]

Whitman's emphasis on departmental coordination was not a trivial concern. He deemed organization an important corollary to specialization; these were, in his words, "companion principles" leading to scientific "progress."[34] His concept of organization was one that permeated not only his institutional plans but his biological research as well. Organization did not just emerge from a combination of cells or faculty members. Nor was it merely responsible for the maintenance of the organism or institution. Organization instead preceded development; it resided latent within the egg as a potentiality of structure passed on through heredity that awaited actualization. Hence, the highest function of organization was "to create and direct," not "to adapt and conform." Organization was thus a guiding force of development and ensured progress along the way. This "precept," Whitman wrote, "is as pertinent to the life of a university as to that of an individual."[35]

Under Whitman's direction, research in the zoology department focused almost exclusively on morphological and developmental questions, with special emphasis on cell lineage studies in developing organisms. As editor of the *Journal of Morphology* and director of the Marine Biological Laboratory (MBL) at Woods Hole from 1888 to 1907, Whitman had a powerful influence over graduate studies in the department. The MBL served as a constant resource for graduate research, and the *Journal of Morphology* was the students' outlet for publication.[36] Of the twenty doctoral dissertations completed by 1902, only three were outside the area of descriptive and experimental biology: two of these were cytological studies, and one addressed the problem of biological variation.[37]

Whitman's own dissertation research on the embryology of the leech *Clepsine* focused on the early stages of cell division that occur prior to germ layer formation. For Whitman, individual development was not merely a consequence of evolutionary history, as Ernst Haeckel argued, but instead contained intrinsic problems that required investigation. During the late 1800s, many investigators looked to Wilhelm Roux's *Entwicklungsmechanik* program as a model for embryological research. Roux divorced embryology from the recapitulation doctrine, arguing that the causes behind individual development could ultimately be explained in physicochemical terms. In a series of experiments reported in 1888, Roux punctured one of the two blastomeres that results from the first division of a fertilized frog egg and then observed the single undisturbed blastomere develop into a half embryo. These experiments seemed to confirm the mosaic theory of development advanced by Roux and August Weismann, in which embryonic growth resulted from a qualitative parceling out during cell division of physical material responsible for cell differentiation. In contrast, Hans Driesch, working with sea urchin eggs at the Naples Zoological Station, achieved normal development from an isolated blastomere at the two-cell stage. His experiments, published in 1891, suggested that development was embedded within the environmental relationships between and among embryonic parts and the surrounding medium. In this instance, the embryo was an individual from the onset, continually adjusting to its changing environment.

Whitman tried to steer the middle course between Roux and Weismann's preformationist stance and Driesch's epigenetic position. In defense of Driesch, Whitman argued that "the organism is not multiplied by cell-division, but rather continued as an individuality through all stages of transformation and sub-division into cells." Organization was already present in the undifferentiated egg. Yet, in viewing organization as the formative process guiding development, Whitman also paid tribute to Weismann's germ plasm doctrine. Organization was itself maintained over generations, preserving the characteristic form of a species through a hereditary component that Whitman identified as "idiosomes." These entities, which preserved individuality and order, linked individual development with the organism's evolutionary past. Indeed, Whitman saw ontogeny as both a product of individual development and evolutionary history—the organism, in Whitman's mind, was first and foremost a historical being.[38]

Whitman's emphasis on organization and historical explanation also filtered into his animal behavior research. During the late 1890s, Whitman returned to the ornithological interests of his childhood: he bred pigeons with the intent of understanding the inheritance and evolution of specific characters. Problems of animal behavior and instinct arose out of these evolutionary studies, and at the time of his death in 1910, Whitman had accumulated a substantial amount of unpublished experimental data and observations.[39] Like his cell lineage studies, Whitman's behavior research was based on comparative analysis viewed from the "common standpoint of phyletic descent." He was primarily concerned with questions of ultimate causation and promoted a broadly conceived "experimental natural history" in which the organism's evolutionary history and heredity were of central concern. Working across a broad range of organisms that included leeches, salamanders, and pigeons, Whitman concluded that the purposive character of behavior was not due "to intelligence" but depended "largely, at least, upon the mechanism of organization. . . . The point of special emphasis here is that instincts are evolved, not improvised, and that their geneology may be as complex and far-reaching as the history of their organic bases."[40] Organization, then, was the underlying factor linking development and behavior, guiding the directional activity of the organism within a loosely bound set of structural constraints. Whitman rejected Lamarckism, favoring an orthogenetic view of evolution in which variation was predetermined along certain lines. He denied that such a view was teleological. Yet, his writings reveal a belief in a purposeful and progressive direction to development at both an ontogenetic and phylogenetic level.[41] His comment to the comparative psychologist Robert M. Yerkes in 1909 that "mnemonic phenomena are in my opinion at the basis of vital phenomena of every order" suggests that Whitman saw development as analogous to learned behavior: the organism, based on its past interactions with the environment, continually restructures its behavior, enabling it to meet future demands.[42]

Whitman's behavior research, based on comparative studies and evolutionary analysis, did not lead to an American discipline of ethology.[43] Wallace Craig and Charles Henry Turner were the only two students to write dissertations that embodied Whitman's ethological approach. But a number of Whitman's students, including Samuel Jackson Holmes and William Morton Wheeler, turned to behavioral studies after they left Chicago.[44] In addressing the relations between instinct and intelligence and the evolution of behavior, Whitman approached the study of behavior from a natural history perspective; he relied almost solely on descriptive and observational techniques. Physiology was, however, rapidly becoming the methodological tool for behavioral analysis at the turn of the century, stimulated by the tropism studies of Jacques Loeb. Loeb, recruited by Whitman in 1892 to develop a program in general physiology at Chicago, centered his approach on immediate causation and the manipulation and control of behavior in the laboratory. His view that phototropic animals were "in reality photometric machines," behavioral responses being nothing more than fixed reactions to external stimuli, struck a discordant note with Whitman who saw behavior as a much more interactive and adaptive response that aided the organism in the course of its evolutionary history.[45] Yet Loeb's physiological approach had a powerful appeal for a new generation of investigators who saw experimentation in the laboratory as being at the forefront of biological research and the key to their own careers. Whitman lamented in 1897 the idea of "a physiologist seriously proclaiming to the world that instinct reduces itself in the last analysis to heliotropism, stereotropism, and the like," but his words fell on deaf ears.[46]

What probably attracted Whitman to Loeb was not his behavior research but his work in physiological morphology.[47] In the early 1890s, Loeb became interested in problems of regeneration as a means for investigating the physicochemical mechanisms controlling growth and animal form. Through the manipulation of external stimuli such as osmotic pressure, contact, or gravitational orientation, Loeb was able to produce such creatures as two-headed hydras, thereby altering the characteristic form of individual organisms. Although Loeb initiated studies in physiological morphology during the 1890s, Charles Manning Child became a leader in the study of regeneration, especially planarian regeneration, from the time of his appointment as an associate in zoology at Chicago in 1896 until his retirement as full professor in 1934. Child shared Loeb's emphasis on mechanistic explanation and analytic technique, but he rejected Loeb's unwillingness to move beyond the immediate practical results of the laboratory to entertain synthetic theories regarding the nature of life.

Loeb's dissatisfaction with his Chicago colleagues and his departure from the university in 1903 suggest an important difference in philosophical approach to the study of biological phenomena. Loeb saw the artificial creation of life as the principal achievement to be accomplished by the experimental biologist, and he grew suspicious of his colleagues' "appeals to either the supernatural or to history as a 'mystical' guide for future social development."[48] Nature did not serve as a normative guide for biological investigation or human action, nor was the biologist's task that of social prophet or healer. For Loeb, the biologist was an engineer, a skillful creator who, within the laboratory, utilized the materials of life to transform nature into artifice, creating products that might perhaps benefit society. In contrast, Whitman, Child, and others looked to biological explanations that brought meaning and purpose to the whole of nature, including humankind. Thus, in his own work on regeneration, Child hoped to construct a theory that would unite not only heredity and development but behavior as well. And Child looked to the behavioral interactions between protoplasm and environment as the unifying force of life in nature, the integrating factor that bound the organism into a functional whole.[49]

In a regenerating planarian, a small piece may re–create its missing part by utilizing already-present cells. In this process, called morphollaxis, some of the fragment's cells change both their shape and their function to re-form as regenerated parts. The phenonmenon of morphollaxis presented an obstacle to Roux and Wiesmann's mosaic theory of development. If, as Roux and Weismann maintained, differentiated cells differed qualitatively, then how could they sometimes dedifferentiate and re-form different parts of the regenerated organism? Child accounted for morphollaxis by suggesting that development went beyond the boundaries of the cell. Like Whitman, he placed importance on the priority of organization in controlling development. Yet, Child's views differed from Whitman's in a significant respect. Whitman, in invoking idiosomes as structural determinants of organization, partially embraced a preformationist stance. Child argued, however, that any predeterminist theory collapsed the problem of individual development into an evolutionary problem where it became "veiled in the mists of the past," relegated "to the field of speculation rather than to that of experiment."[50] Child sought instead a strictly physiological account of development, one with a reference frame that encompassed "the development, maintenance, and relation to environment of the individual organism in a protoplasm of specific hereditary constitution, rather than . . . the evolution of this specific constitution."[51] He saw form as superfluous, an epiphenomenon created by the functional activity of the organism. "The organism," Child wrote, "is primarily a dynamic or functional complex and the process of morphogenesis is merely an incident, or, in other words, structure is a visible by-product of these activities."[52] Any doctrine such as Weismann's that invoked morphological entities to explain the emergence of organic form assumed precisely what it set out to explain. Organic structure, Child maintained, was not the result of a substance's constitution but arose instead from "the relations and interactions of the elements in a given environment."[53] Hence, for Child, function, not structure, was the ultimate determinant of organization.

In his persistent search for an integrating factor responsible for the unity and order of the individual organism, Child had, by 1911, postulated the existence of physiological gradients within the organism. He suggested that areas of increased metabolic activity would occur in the undifferentiated protoplasm in response to environmental stimuli. These areas of excitation would become imprinted in the protoplasm, establishing a higher metabolic rate than the region farthest from the source. According to Child, the major gradient of development originated along the anterior-posterior axis of the organism; thus, the apical region of the head dominated and controlled differentiation in the subordinate parts. Child cited experimental work on anesthetics as evidence for the existence of metabolic gradients. He subjected regenerating *Planaria* fragments to a wide variety of metabolic poisons, reasoning that the most physiologically active portions of the organism would take in the poison more rapidly and therefore die more quickly. Over a lifetime of work he accumulated evidence that *Planaria* are indeed metabolically graded.

Child initially began his graduate training in Wilhelm Wundt's laboratory of experimental psychology at Leipzig, and his early interest in psychology is reflected in the central importance that behavior played in his developmental theory. He looked favorably on the attempt by the German biologist Richard Semon to "interpret the phenomena of heredity, habit, and memory on a common basis."[54] Building on the analogy between heredity and memory advanced by Semon, Child argued that the physiological gradient "represents, in short, a protoplasmic memory, and so constitutes the first step in the education of the protoplasm concerned." These gradients, once established, influenced all later behavioral responses between protoplasm and environment.[55]

Child's insistence that the organism was "inexplicable without environment" mirrored, in many respects, the pragmatic philosophy advanced by his colleagues such as John Dewey and George Herbert Mead. In fact, Child pointed out that the organism-environment dualism at the base of predeterminist theories had led to "sterility" in "various fields of biological thought" and had created a similar effect in philosophy.[56] Dewey and other Chicago pragmatists rejected metaphysical dualisms such as mind/matter, subject/object, and stimulus/response. These entities were instead integrally related through the organism's activities. In his famous 1896 paper, "The Reflex Arc Concept," for example, Dewey pointed out that stimulus and response were not distinct but were instead parts of a coordinated whole that only took on meaning with reference to the experiential act.[57] In Dewey's functionalist approach, the individual could only be defined with respect to its environmental interactions. New environmental conditions called forth new responses in the individual until an equilibration between organism and environment was once again reached. What is significant for this discussion is that Dewey saw education as a "continuing reconstruction of experience in the light of past experience and present problems."[58] Dewey put this philosophy into practice in his Laboratory School at Chicago. He espoused a pedagodgy built on the premise that "a possibility of continuing progress is opened up by the fact that in learning one act, methods are developed good for use in other situations."[59]

Both Child and Dewey saw behavior as a process of life adjustment whereby the organism structured its environment to meet new demands. Their perspective added a different dimension to the environmental determinism of Salisbury's physiographic program: namely, the organism as a modifier of its environment. The relationship between environment and organism was not unidirectional but interactive. Herbert Spencer Jennings's experimental work on invertebrate behavior, which was an essentially "biological" approach, focused on internal physiological states and individual variability and thus had a greater appeal for Child than did Loeb's tropism studies. Loeb, by interpreting behavior as a mechanical response to external stimuli, placed the cause of behavior in the animal's external environment. Jennings thought otherwise. As an undergraduate at the University of Michigan, Jennings was a disciple of Dewey. In 1896, Jennings completed his dissertation in zoology at Harvard.[60] Fostered by postdoctoral research conducted at the University of Jena under Max Verworn, Jennings's interests turned to the behavioral reactions of protozoa. Unlike Loeb, Jennings found that infusioria, such as *Paramecium*, did exhibit a great deal of flexibility in response to given stimuli. Reactions could not always be "determined by the direct action of a localized stimulus" but often depended on the internal, physiological state of the organism.[61] Thus an organism responded to external stimuli by a series of "trial and error" movements, changing its direction until the disturbance disappeared. Behavior was a regulatory process of adaptation whereby "the organism tends to find conditions favorable to its life processes and to retain them."[62] Because Jennings emphasized the physiological conditions of the organism as a causal factor in any behavioral response, his analysis, as compared to Loeb's, required a much more detailed understanding of the organism under investigation. What, for example, were the natural environmental conditions in which the organism was found; in short, what was the organism's life history? In analyzing the interactions between organism and environment, Jennings, like Child, placed the organism at the forefront of the analysis.

Child's physiological theory of development and his rejection of any biological theory that smacked of preformation placed him at odds with the growing separation between the study of heredity and development in American biology during the first two decades of the twentieth century. At the turn of the century, inheritance included the two areas that we today call genetics and embryology. However, with the rediscovery of Mendel's work in 1900 and the eventual resolution of Mendelism with a physical theory of inheritance by Thomas Hunt Morgan and others during the 1910s, the field of inheritance took on an extremely narrow scope.[63] Wilhelm Johannsen's 1911 paper, "The

Genotype Conception of Heredity," revealed increasing separation between the study of heredity and development. Under the "genotype conception," heredity was to be defined as "the presence of identical genes in ancestors and descendants."[64] Genetics would be limited to the study of character transmission, as geneticists traced characters from one generation to the next and isolated them at a physical location on the chromosome. The historical expression of those characters within the individual, that is, the phenotype, was relegated to the study of embryology.

Child vehemently opposed the attempts to define heredity as transmission genetics alone and to isolate the chromosome as the fundamental unit of analysis. Chromosomes were simply morphological entities, by-products of functional processes taking place within the organism. Chromosome mapping was, according to Child, mere descriptive anatomy. Meanwhile, the underlying processes that made the chromosome and were responsible for organization escaped notice. Heredity, Child argued, was not the "genetic history of the germ plasm or its determinants or unit characters" but was the "capacity of a physiologically or physically isolated part for regulation." "Wherever reproduction of any kind, whether of parts or of wholes, occurs," Child maintained, "there we have also to do with heredity."[65]

In Child's view, the uniformity of individuals within a given species was the result of two factors: the hereditary constitution of protoplasm and the continuity of the organism's developmental environment. Mendelism, which by 1924 had "modified in many ways . . . conceptions of the mechanisms of inheritance . . . ," nevertheless still had "to a large extent ignored the problem of the mechanisms by which particular hereditary potentialities are realized as characteristics of particular individuals."[66] Geneticists had focused on the stable nature of protoplasmic substance while neglecting to consider the constant environmental conditions to which a developing organism was subject—they had ignored the "education" of protoplasm. Alter this uniform environment, Child argued, and the hereditary potentialities of protoplasm that lie latent within a given individual become expressed.

Child's skepticism with respect to Mendelian genetics and his emphasis on the physiology of development was also shared by his colleague and department head Frank R. Lillie. Lillie received his doctorate in zoology at Chicago in 1894 under Whitman and returned as an assistant professor of embryology in 1900. Unquestionably Whitman's chosen successor, Lillie took over the daily administrative affairs of the department. In 1910, he became official chairman of the department, a position he held until his appointment as dean of the Division of Biological Sciences in 1931. His marriage to Frances Crane, sister of the Chicago plumbing magnate Charles R. Crane, placed him in the social circles of wealthy elite, such as the Rockefellers and Carnegies, individuals who were themselves trustees and sponsors of the various foundation boards funding science during this period. These connections proved advantageous for Lillie on many occasions. When he assumed directorship of the MBL in 1908, for example, he was able to ensure that the MBL maintained its role as a preeminent research institution through the donations of his brother-in-law for a new laboratory.[67]

Like Child, Lillie argued that "an immense part of what we call inheritance is inheritance of environment only, that is, repetition of similar developmental processes under similar conditions."[68] Physical characters were not units in a Mendelian sense, but were instead products of physiological processes taking place within the organism. Process, not structure, was the starting point for any analysis of heredity and development. And developmental processes were themselves interactive. "The developing embryo," Lillie wrote, "is not merely a unit on which an extra-organic environment operates, but it is a living mosaic, each element of which may conceivably enter into the development of any other in the sense of being a factor in the process."[69]

A vision of the organism and its development came into focus at Chicago during the early 1900s. This outlook, expressed perhaps most succinctly by Child and Lillie, had a marked impact on the kind of support that animal ecology research subsequently received. Three features of this perspective stand out. The first is Child's and Lillie's refusal to separate heredity from development, as evidenced by their opposition to Mendelian genetics. This opposition continued well into the 1920s. When, for example, a position in genetics came open in the department in 1924, four candidates were under consideration: J. A. Detlefsen, Sewall Wright, R. E. Clausen, and H. J. Muller. "As regards Muller," Child wrote, "I cannot help feeling that the Morgan crowd is somewhat adrift in speculative predeterminism. Personally, I should not be inclined to pick him."[70] In keeping with Child's opinion, the department hired Sewall Wright, whose own work in physiological genetics proved much more amenable to the developmental outlook of the department.

When Child branded Morgan's work "speculative predeterminism," he was suggesting the extent to which Mendelian genetics had abandoned the study of environment in development. If everything was predetermined in the chromosomes, as the physical theory of heredity outlined by Morgan and others implied, then the interaction between developing embryo and environment was superfluous. Yet Child placed that interaction at the very center of the developmental process. Both Lillie's and Child's embryological research focused on the immediate environmental determinants governing animal form. By advocating a physiology of development, they departed from their mentor, Whitman. While Whitman stressed embryological study in light of the individual's phylogeny, Child and Lillie downplayed the significance of heredity and evolution. What significance did this have for ecology, which is, first and foremost, a study of environmental relationships? At Chicago, animal ecology developed quite independent of the study of heredity as defined by Mendelian geneticists; the study of animal distribution would center not on transmission of characters, not on the historical relationships and variation among species within a given area, but on the behavioral similarities between unrelated organisms that gather in an area as a consequence of immediate environmental conditions. From an ecological perspective, the interaction between organism and environment was a key factor in shaping the form of animal communities. As the next section shows, Child had a significant impact in pushing ecological research in the department toward this physiological direction.

The opposition to Mendelian genetics voiced by Child and Lillie also points to other important elements that became a part of the Chicago program of animal ecology research. As Peter Bowler has argued, Mendelian genetics struck the deathblow to evolutionary theories modeled on a developmental philosophy of individual growth guided toward purposeful ends. If variation arose only through genetic mutation, then evolution was indeed directionless and nonprogressive—a chance process with no inherent unity or underlying plan. Yet the embryological focus of departmental research at Chicago underscored a developmental metaphor patterned on the growing embryo as it proceeded through the various stages from egg to adult. Here was a process both goal directed and progressive that could be applied to evolution as well. Ontogeny and phylogeny, development and heredity, were part of the same historical unfolding. In ecology, the most evident pattern of change, especially in the early 1900s, was succession. Pioneer plant and animal communities occupy barren ground, and a sequence of stages occurs until the mature climax condition is approached. Succession is distinctly a developmental metaphor patterned on individual growth. One begins with barren ground just as one begins with the unfertilized egg, and development proceeds to the climax community just as it proceeds to the adult individual. Both are part of the same Spencerian evolutionary pattern characterized by growth from the homogeneous to the heterogeneous, from the simple to the complex. Hence, in embracing a developmental picture that was goal directed and progressive, embryological research at Chicago lent credence to one of ecology's earliest cherished principles.

Succession, however, as understood by Cowles in his plant ecology research, was driven primarily by physiographic change. Cowles, as Salisbury hoped, studied the effects of the physical environment on organic life. But this rather blind environmental determinism left little room for the organism as active participant in its future, nor did the "speculative predeterminism" of Mendelian genetics, which Child disdained. Just as Dewey saw learning as an adaptive process in which the individual utilized past experiences with its surroundings to respond to present and future conditions, so Child saw the form of the developing embryo structured by the continually changing organism-environment relationship. The organism did not just mechanically unfold irrespective of surrounding conditions, nor was development driven strictly by the physical environment; rather, the organism and environment were constituents of the same developmental process. As the undifferentiated egg reacted to its environmental surroundings, physiological gradients were established which, in turn, created a new environmental complex of intercellular relations that guided and directed the future course of development. The organism, through its behavior, conditioned the environment, which in turn allowed the succession of other developmental stages, whether at the intercellular or interorganismic level. This focus on behavior and the organism as a causal factor in development was to become a central element of animal ecology research at Chicago from its inception to the departure of Warder Clyde Allee in 1950. How these ideas initially coalesced into a program of animal ecology research is the subject of the next two sections.

Distributional Patterns

For Whitman and the zoology department, 1899 was a particularly trying year. The loss of William Morton Wheeler and Shosaburo Watase to other institutions reduced the staff to just Whitman, Child, and Edwin O. Jordan, an assistant professor in bacteriology. Whitman complained bitterly to Harper of the situation. Years of financial promises had come to nothing. The death in that same year of George Bauer, a professor in paleontology, left three positions vacant, but Harper was only willing to fund two. "Our *de*velopment," Whitman bitterly mused to Harper, "is thus of a curious sort, much like the *en*velopment of last century philosophy; not from less to more but from less to less *infinitum*."[71] The situation was, however, not a complete loss.

Figure 2.2. The Department of Zoology, 1901. *Standing, left to right,* Frank R. Lillie, Charles Manning Child, and Charles Otis Whitman. Charles Benedict Davenport is standing on the far right. Charles C. Adams is seated in the second row, third from the left. Courtesy of the University of Chicago Archives.

Whitman used one position to retrieve his protégé Lillie from Vassar College. The other position, although first offered to Thomas Montgomery, went to Charles B. Davenport, an Instructor in Zoology at Harvard University (fig. 2.2).

Davenport received his Ph.D. at Harvard in 1892 in a department in which "research was directed almost exclusively along morphological lines." Captivated by the new experimental methodology embodied in Roux's *Entwicklungsmechanik* program, physiological morphology studies, and invertebrate behavior research, Davenport published a two-volume treatise, *Experimental Morphology,* in 1898. The purpose of the book was to analyze the external environmental causes responsible for development. The first volume considered the effects of such agents as chemical substances, water, density of the medium, molar agents, gravity, electricity, light, and heat on living protoplasm, while the second examined the effects of the same agents on growth. During the 1890s, Davenport also taught a class on high school zoology each summer to teachers at the Biological Laboratory of the Brooklyn Institute of Arts and Sciences at Cold Spring Harbor, New York. Cold Spring Harbor, a fjordlike inlet that opens into Long Island Sound, is situated about twenty-five miles northeast of New York City. The beach along Cold Spring presented conditions similar to those of the Indiana Dunes, and Davenport reflected that his work on experimental morphology, "which dealt with the effects of environmental conditions upon protoplasm and growth," together with the "environment of Cold Spring Harbor" lured him "into ecology." He might also have included Chicago and H. C. Cowles in his list of environmental influences.[72]

In the summer of 1900, after Davenport's first year of teaching at Chicago, he and Cowles led a class of students to the Cumberland Mountains of eastern Tennessee en route to Cold Spring Harbor, where Cowles had been invited by Davenport to head botanical instruction at the Brooklyn Institute's summer school. Davenport and Cowles continued to lead joint field ventures until Davenport's departure from Chicago for a permanent position as director of the Cold Spring Harbor station for the study of experimental evolution in 1904. Their excursions included not only New York and Tennessee but also the Gulf Coast of Mississippi, where they led a delegate of students in 1902.[73]

These joint associations, together with his interests in experimental morphology, stimulated Davenport to initiate research, published in 1903, on the animal ecology of Cold Spring Harbor. Working on the sand spit that separates the inner harbor from the outer waterway opening into Long Island Sound, Davenport divided the beach into three regions (the submerged zone, the lower beach, and the upper beach) and noted the distribution of fauna that occurred in each. He then compared the distribution of fauna to that found along the shoreline of Lake Michigan. Although the species were different, Davenport remarked that within the submerged zone, both beaches possessed sessile, crawling, and swimming fauna. Furthermore, when comparing the lower and upper beaches of the two environments, similar forms of carrion flies, carrion beetles, robber flies, tiger beetles, and white grasshoppers were found. Davenport concluded that "the fauna of a point is, within limits, determined rather by environmental conditions than by the geographic position of the point."[74] Just as the embryologist could examine the form and development of an individual organism by reference to either the immediate environmental relationships between and among cells and their environment or the organism's phylogenetic past, so too could one similarly separate the determinant factors of animal distributions into proximate causes, based on the animal's abiotic and biotic surroundings, and ultimate causes, centered on historical relationships linked by descent.

Having surveyed the faunal distribution, Davenport focused on the location and behaviors of three species of Collembola (a small, wingless insect) on the Cold Spring beach. He observed a pattern to the movements of the Collembola correlated with the rising of the tide. As the tide rose, the Collembola leaped into the air, and on touching the ground, whirled so as to bury themselves in the sand to a depth of six to nine inches. When the water receded, they rose to the surface, running up stones and moving into the wind. Davenport explained each of these movements on the basis of behavioral responses to oxygen, contact, light, gravity, moisture, and touch. For example, the descent of Collembola into sand was the result of a positive reaction to contact and moisture. Furthermore, by comparing the responses of the beach Collembola to other species within the same order, Davenport concluded that these reactions were properties of the entire order. The instincts of the beach Collembola were not adaptations to a beach habitat, because they existed independent of, and prior to, their beach existence. Habitat had not determined instincts; instead, the Collembola's instincts had "determined their habitat."[75]

Davenport used this evidence to support his "theory of segregation in the fittest environment." According to this hypothesis, adaptation was "not due to a selection of structure fitting a given environment, but, on the contrary, a selection of an environment fitting a given structure."[76] The organism, already possessing certain characteristics, selected the habitat most congenial to its particular organization. Davenport's theory was almost identical to Jennings's characterization of behavior as a trial-and-error response where the organism adjusted itself to the surrounding environment until physiological equilibrium was reached. Both were in fact modified versions of the theory of organic selection advanced by James Mark Baldwin, Henry Fairfield Osborn, and Conway Lloyd Morgan in the 1890s. Organic selection, or the Baldwin effect as it later became known, postulated that an organism might learn behaviors beneficial to its survival. These learned behaviors would enable the organism to perpetuate itself in a new environment, and over the course of many generations, chance variations coincident with these acquired behaviors would be selected for. Eventually the behavior pattern would have a hereditary foundation. Organic selection enabled the organism through its behavior to exert some purposeful control over its development and evolution without recourse to the inheritance of acquired characteristics.[77] The theory, as advanced by Davenport, also lent support to Hugo de Vries's mutation theory, and to theories of nonadaptive evolution in general, because adaptation was no longer a condition of selection in the Darwinian sense. Rather, the organism possessed characteristics that were conducive to its survival in a particular environment, even though these variations may have arisen by chance in different environmental conditions where they had no adaptive significance. In a period when most biologists regarded selection as a destructive rather than a creative force, Davenport's theory had popular appeal.

Davenport never followed up on his early foray into the field of animal ecology. After a trip in the early 1900s to England, where he visited Francis Galton, Karl Pearson, and Walter Weldon, pioneers in the study of biometry, Davenport became committed to the "quantitative study of Evolution."[78] His courses at Chicago, which included experimental and statistical zoology, experimental evolution, variation, and zoogeography, reflect his growing concerns with heredity and variation as experimental inroads into the problem of evolution. In fact, in his course on zoogeography, little of the behavioral approach to the study of animal distribution present in his Cold Spring Harbor study was apparent. Instead, the course covered the "principles of the geographical distribution of animals, with special reference to evolution."[79]

Two approaches to the study of animal distribution—one evolutionary, the other physiological—existed side by side at the University of Chicago in the early 1900s. Charles C. Adams and Victor E. Shelford, both classmates in the zoology department around the turn of the century, were the respective representatives of these two divergent outlooks by 1908. Adams followed Davenport from Harvard to Chicago in 1899. In 1903, he took a position as curator of the Natural History Museum of the University of Michigan, where he led state biological surveys while working on his Ph.D., which he received from Chicago in 1908.[80] Adams's extensive experience with museums and state natural history surveys led him to emphasize a methodological approach to animal ecology that was largely descriptive and observational. Through the encouragment of Davenport and Whitman, he also paid heed to taxonomic and evolutionary problems.

In the summer of 1900, Adams accompanied Davenport and Cowles to the Cumberland mountains of Tennessee. This location played a central role in his early research, which sought to understand the impact of baseleveling and changing physical conditions on the migration and dispersal patterns of North American fauna. He postulated that the southeastern United States served as the center of origin for eastern North American flora and fauna in the postglacial period.

Through a physiographic analysis of the region, with the appropriate acknowledgments to Cowles and Salisbury, Adams uncovered what he argued were the three main highways for faunal migrations: the Mississippi valley, the Coastal plain, and the southern Appalachians.

Support for hypothesized lines of dispersal, Adams argued, could be derived from a study of variation. His own work on gasteropod shells in the Tennessee River system indicated that the oldest, least variable forms were confined to the headwaters and a progression toward increased variability in the number of spines occured as one moved downstream. Paying tribute to Osborn's law of adaptive radiation, Adams suggested that "the life of a region" is "constantly diverging or radiating from its original home and parental stock, encountering new conditions of environment and becoming modified in both habits and structure."[81] The study of animal distribution, Adams maintained, must proceed from a "dynamic and genetic standpoint." By this, Adams meant that one must first begin with the problem of faunal origins and from there unearth the physiographic processes and biotic conditions by which life radiated out from a given region. Adams worked within a framework of geologic time; his successional model took place on an evolutionary stage where species and their descendants provided the continuity between acts. Adams, in fact, distinguished between two versions of succession: "the adaptational one, in which the ecological aspect is prominent, and the hereditary one, in which the taxonomic or hereditary aspect receives emphasis."[82] He favored a dendritic or treelike classification, as opposed to a horizontal one based on ecological relations, to preserve similarities between animal forms due to common descent.

Adams's genetic approach to animal ecology received little support in the department. When Davenport left in 1904, Adams turned to Whitman as an adviser, in whom he found a sympathetic ear. But the department's future now resided in the hands of Lillie and investigators such as Child who embraced a developmental perspective centered on physiological processes and proximate environmental causes as determinants of animal form. Shelford was the individual who took this vision and shaped it into a program of animal ecology research, one that mirrored the experimental morphology approach abandoned by Davenport in his Cold Spring Harbor study.

Defining a Field

Shelford entered the zoology department as an undergraduate in 1901, after completing two years of undergraduate education at West Virginia

University. The majority of his courses were taken under Cowles, Davenport, and Child. His dissertation, "Life Histories and Larval Habits of the Tiger Beetles," completed in 1907, combined Davenport's focus on animal habits and physiological response with Cowles's research on plant succession to arrive at a distributional model of tiger beetles on the Indiana dunes. Child and Whitman also had an influential role in shaping Shelford's ecological studies. The emphasis Whitman placed on the study of life histories is evident in Shelford's dissertation, and Shelford later thanked Whitman for "encouraging the study of natural history." Yet Shelford ultimately rejected Whitman's evolutionary focus in defining a field of research. Although he often referred to problems of evolution in his dissertation, such references were notably absent in his programmatic statements that outlined the study of animal ecology. Instead, Shelford incorporated the physiological outlook of Child into his own methodological approach.[83]

In his dissertation, Shelford sought a causal explanation for the particular distribution of tiger beetles found on the Lake Michigan shore. By examining the life cycle of tiger beetles in the field, he noticed that although the range of the adults varied, the distribution of the larvae was correlated with particular soil conditions. He placed adults of various species in cages that differed in soil composition and slope and noted the number of larvae that developed in each case. Each species had an optimal range of soil conditions favoring development, which suggested that the egg-laying habits of the tiger beetles were the limiting factors governing their distribution. Reproductive behavior was, in this case, the least plastic response and therefore placed the most stringent constraints on the environment in which the beetle could survive.

Shelford had introduced an experimental component into the study of animal distribution. By manipulating soil conditions in the laboratory, he was able to ascertain the environmental conditions influencing reproductive behavior and from this, determine the causal factors of animal distribution in the field. Furthermore, by using Cowles's classification of plant societies on the dunes, Shelford could also correlate beetle distribution with plant succession based on underlying soil conditions. For example, *C. sexguttata,* which deposits its eggs in a clay-humus mixture and in partially shaded conditions, is most abundant in the white oak–red oak–hickory association. But as succession proceeded to the more mesophytic beech-maple forest, the increased humus and shade would be unfavorable to the egg-laying habits of this species, and it would gradually be replaced by another form.[84]

After receiving his degree in 1907, Shelford directed his attention toward establishing a respected program of research. He remained at

Chicago as an associate and then instructor on the zoology faculty until 1914. In 1908, the field zoology course at the senior level was replaced by animal ecology, which focused on the "breeding habits, and interrelations of the forms that make up the animal societies of the local faunas, and the relations of these societies to physiographic and environmental processes–such as erosion, deposition, and plant succession." By 1910, a whole sequence of courses on behavior and ecology was offered. These included field zoology, animal ecology, geographic zoology, animal behavior, and a graduate seminar on topics in ecology conducted in conjunction with the botany department.[85]

With the publication of "Physiological Animal Geography" in 1911, Shelford's intentions for animal ecology became known. In this paper, Shelford elicited a new line of inquiry. Traditional faunistic animal geography, which included Adams's genetic approach, was inadequate because of its close alliance with evolution. In fact, Shelford argued, "the subject is even more strongly committed to speculative evolution than any other phases of biological science."[86] Shelford wanted to establish instead an analysis of animal distribution from the standpoint of physiological inquiry, an analysis freed of descriptive morphology and speculative evolution. Evolution, in Shelford's eyes, was ecology's downfall. Emphatic in his position, Shelford reminded his readers that "regardless of widespread ideas to the contrary, ecology or ethology belongs primarily to the physiological point of view and is therefore outside of the range of criticism from the point of view of evolution or the current germ plasm doctrine. Its frequent confusion with various branches of evolutionary speculation, such as mimicry, structural adaptation, etc., is one of the commonest errors of recent writers and has been chiefly responsible for such prejudices as may possibly exist."[87]

Shelford was especially critical of August Weismann's germ plasm doctrine because it emphasized the fixity of hereditary characters in Darwinian evolution while ignoring the "experimental study of response." He disliked the idea that the germ plasm had a separate life from the soma, that it existed "completely insulated from the environment" over the course of generations. Such a theory, Shelford recognized, confined the study of heredity to the chromosome and character transmission and eliminated any discussion of environmental influence. Zoological work was being judged with respect to its bearing on the problem of heredity narrowly defined, while the study of somatic characters and individual development took on "secondary importance." Shelford adopted the same critical attitude toward Mendelian genetics as Lillie and Child.[88]

The morphological bias of the "germ plasm criterion" was, Shelford maintained, giving way to that of the physiologist. Because animal physiology had previously "been isolated in medical schools," its influence on genetics, faunistics, and morphology, had been slight. This was, however, changing. Shelford was himself trained in an institution in which biology developed uninfluenced by the demands of medicine and that allowed a curriculum in general physiology to be established, at least for a time.[89] He had taken general and theoretical physiology in the Department of Physiology and received a further push in this direction through Child and Lillie's own focus on the physiology of development. Experimental morphology sought a causal analysis of individual organic form through a study of protoplasmic response to various environmental factors. Shelford extended the level of analysis beyond cells to include animals living in a given region. He rejected attempts to restrict the problem of animal distribution to the study of heredity and evolution, just as Child and Lillie refused to defer embryological research to problems of phylogeny. The "study of physiological animal geography," Shelford insisted, "may be conducted independently of the problems of evolution. It does not need to be concerned with centers of origin, or paths of dispersal. . . .[It is] concerned solely with the physiological relations of animals to natural environments."[90]

Shelford coined the word, *mores,* or rather, shifted meanings from human to animal, to refer to the behavioral and physiological characteristics of a group of organisms. In plants, structures are relatively plastic and can be used as an index of physiological response. Animal form, however, is not so malleable and gives little indication of the surrounding environmental conditions. The animal ecologist's counterpart to plant form was, according to Shelford, behavior. "The behavior and general mode of life of animals," Shelford wrote, "are the superficial equivalent of the structural phenomena in the vegetative parts of plants." Both were "convenient indices of physiological conditions within the organism."[91] Traditional taxonomy had failed to reveal ecological relationships among animals because it classified organisms according to "structural adaptations." By this, Shelford meant morphological characters as opposed to animal activity or behavior. The taxonomic classification of tiger beetles based on structural adaptations in Shelford's dissertation research, for instance, gave no indication of the differing *habits* of these beetles in specific environmental conditions. Shelford wanted instead a horizontal method of classification that related taxonomically diverse groups on the basis of similar mores. To determine the important environmental factors governing animal distributions, therefore, one had first to arrive at a detailed understanding of the physiological life histories of the different organisms present in a given community.

Shelford's first point of entry into the domain of animal ecology was to gather information on the behaviors, or mores, of individual organisms within a single species or genus. This approach fell under the heading of individual or aggregate ecology (also referred to as autecology), and here the laboratory served as the organism's milieu.[92] Shelford believed that behavior was a good index of physiological conditions and gave important clues to the environmental factors influencing where the organism was found. By analyzing behavioral responses to physical or chemical factors deemed important in the environment, the researcher might find the agents to which the organism was most sensitive. These agents, concomitant with the specific behavior, would be the limiting factor governing the animal's distribution. The underlying assumption was the same as that employed by Child in his regeneration experiments, namely, functional response determines organizational pattern.

Warder Clyde Allee's dissertation on isopod behavior typifies the research strategy employed in this work on the physiological life histories of individual organisms. Most of Allee's courses when he entered the zoology department in 1908 were taught by Shelford and Child. Although Allee did take a few courses in botany, none were with Cowles; instead, all of his botany classes were in plant physiology and chemistry.[93]

To investigate the causal factors determining isopod distribution, Allee collected isopods from two different locations: young streams and older ponds. Since all the isopods were the same species, differences in behavior could not be ascribed to heredity; rather, they were a consequence of the environment. Allee first placed the animals in an artificially produced current and noted their reactions. The stream isopods showed a high degree of positive reactions, that is, they would move in the direction of the current, while the pond isopods gave a weak positive response. Furthermore, the number of positive responses decreased in stream isopods collected during the breeding season. Allee postulated that the degree of positive rheotaxis depended on the metabolic rate of the animals. If the animal's metabolism was lowered by breeding activity, this would account for the decrease in positive response. The relation between breeding and rheotaxis affected isopod distribution in the same way that breeding habits determined tiger beetle distribution in Shelford's research. Even though the normal rheotactic response of isopods was sufficient to maintain them in strong currents, they were rarely found in such places. Instead, the weakened rheotactic response of the isopods during the breeding season limited them to streams where protected places could be found.

The difference in rheotactic response between the stream and pond isopods was also indicative of varying metabolic rates. Since these differences depended on environment, Allee suggested that oxygen or carbon dioxide content of the water might be the causal factor that accounted for behavioral differences. Oxygen and carbon dioxide concentrations clearly affected metabolism, and Allee found significant differences in concentrations between pond and stream water. To test this hypothesis, he subjected stream isopods to environmental stimuli that depressed metabolism and recorded the rheotactic response. Allee adopted the same experimental techniques as those developed by Child in his work on anesthetics and physiological gradients and applied them to an ecological problem. Low oxygen, chloretone, potassium cyanide, low temperature, carbon dioxide, and starvation—the same depressing agents used by Child—all decreased the positive response in stream forms. Moreover, pond isopods exposed to high oxygen concentrations, caffeine, and high temperatures exhibited a high degree of positive rheotaxis characteristic of stream isopods. They did not, furthermore, exhibit a return to the normal pond response even when the experiment continued over six months. The behavioral differences were thus, Allee reasoned, dependent on environmental rather than hereditary factors. Once again, taxonomic classification failed to give any indication of the physiological differences and environmental influences that separated the two groups. Heredity, as defined by geneticists, was simply inadequate for understanding ecological relationships.[94]

Allee's isopod research is but one example of the attempt, under Shelford's directive, to explain animal distribution on the basis of physiological response to environmental stimuli.[95] Still, the analysis was not complete. By amassing an assemblage of physiological life histories of different organisms within a community, Shelford hoped to arrive at an understanding of the ecological similarity of different species occupying the same habitat. Yet this aspect of his program was much harder to fulfill. It was difficult enough to analyze experimentally the behaviors of a single species, let alone an entire association of animals within a given area. Shelford had shifted levels from the individual to the community. But the methods employed in community or associational ecology (also referred to as synecology) were largely descriptive, and considerable attention was devoted to the problem of ecological succession. Although Shelford originally viewed the methods of autecology and synecology as operating in tandem, by the 1910s, individual and community ecology were developing along separate lines.

In the domain of community ecology, Shelford enthusiastically cited a paper read in 1907 by his classmate Wallace Craig to the Association of American Geographers at Chicago as a model for research. Craig entered graduate school in the zoology department in 1901 and studied the behavior of pigeons under Whitman, for which he received his Ph.D. in 1908.[96] Between 1905 and 1907, he taught school in Valley City, North Dakota and translated his experiences on the plains into an article on North Dakota life. Three features characterized North Dakota's physical landscape: flatness, aridity, and severe winter. Craig analyzed the influence of each of these environmental factors on plant, bird, mammalian, Indian, and pioneer life and noted the similarity in behavior among these diverse organisms. The wide expanse of the prairie, for instance, favored the development of large plant associations and extensive animal herds such as the bison. This tendency toward social aggregation and organization, influenced by the "traversability and uniformity of the plains," was also evident in human life. Although the actual number of people living on the plains was small, Craig saw a "geographic unity, a social solidarity, and a political discipline" that did not exist among people living in mountainous or forest terrain. The severe winters also fostered increased sociality, as evidenced by animals herding together to keep each other warm or by the hospitality of the North Dakota farmer during blizzards.[97]

Craig's analysis was appealing to Shelford because it bound taxonomically diverse groups of organisms within a region together on the basis of behavioral similarities. An ecological agreement among the different kinds of organic life could be discerned, an agreement due to the presence of common environmental conditions. Furthermore, Craig's study showed the intimate relation of ecology to human geography. Geographers, sociologists, and psychologists had failed to see the significance of ecology for their disciplines, Shelford argued, because they compared structural adaptations in plants and animals to cultural adaptations in human societies. The appropriate point of comparison was, however, animal mores. The ground nests of birds, the underground burrows of prairie dogs, and the sod-house shelters of the pioneers, for example, all reflected the same behavioral response to arid conditions. In examining the behavioral interactions among individuals within a species, Craig had undertaken an investigation into what Shelford called interphysiology or interpsychology, a term he borrowed from the French sociologist Gabriel Tarde. Interphysiology was essentially a study in social life, but it addressed only one element of community integration—the bonds that unite individuals of the same species into a functional whole. The question of how different species of animals were integrated within the community was left untouched.

This was the subject of intermores physiology, or the study of the interactions between organisms "antagonistic in behavior and habits." Organisms were also linked through such processes as competition and feeding relationships, and this is where ecology, Shelford argued, "comes into contact with theories of natural selection, adaptation, and mimicry."[98]

Shelford did attempt to analyze the physiological agreement of animal species within a given community. He collected the most abundant species of a rapids community, for instance, and determined the reactions of each species to current, bottom, and light. All species showed a similar preference for hard bottom and a positive response to strong currents, while they differed in their reactions to light. These differences could be accounted for by reference to the vertical stratum in which the species lived, that is, under stones, among stones, or on stones.[99] These experiments lent support to the view that animals gathered in a region on the basis of similar physiological responses to environmental conditions. Such an approach, however, yielded little information on intermores physiology, on the interaction among species within a community. This is where succession served as a powerful descriptive model because it directed one's attention to organic interactions that precipitated developmental change.

Shelford's most significant contribution at Chicago to community ecology and the problem of succession was a five-part series on the ecological succession of aquatic and terrestrial communities in the Chicago region that was expanded into a 1913 monograph, *Animal Communities in Temperate America*. In this work, Shelford began with the physiographic method employed by Cowles. He found an ecological succession of fishes coincident with the stream's physiographic age: similar species of fish were found in both young streams and the headwaters of the oldest streams. For Shelford, however, further analysis was necessary. Physiography was simply an ecological tool for placing the organism in its physical environment; it was not a causal explanation of ecological succession. This is where animal ecology at Chicago departed from plant ecology's physiographic emphasis. Whereas Cowles saw changes in the physical landscape as the driving force behind plant succession, Shelford searched for an explanation of succession that was derived first and foremost from knowledge of the organism. Living beings interacted with and transformed their physical surroundings; they were not mere pawns of geologic change.[100]

Once the animal had been placed in its environment through physiographic methods, Shelford hoped to determine the causes of succession. His studies on fish succession in the Chicago region were situated in an area south of Lake Michigan, where a series of ponds was arranged in a horizontal sequence according to geological age: the ponds closest to the lake were the youngest in the series. The horizontal succession of fish communities in these ponds was representative of the biotic succession that resulted in a single pond over time. Shelford compared nutrient levels, oxygen content, food species, and competition in the different ponds to ascertain the important factors governing fish succession. None, however, were as important as the fish's breeding habits. As the ponds aged, organic material accumulated, causing an increase in vegetation, a decrease in the water's oxygen content, and a change in the bottom substrate, all of which had an adverse affect on the fish's breeding behavior. The fish species would be replaced by others as the environment became detrimental to their breeding habits. The succession of fish communities was thus a consequence of the "succession of breeding conditions and breeding mores." The organism, through its behavior, found conditions conducive to its survival. When these conditions changed, either through the organism's own activities, through the influence of other organisms, or through physical changes in the landscape, a species with different needs would gradually take its place, and the whole assemblage of organisms would change.[101]

Shelford left Chicago in 1914 for a position as assistant professor in zoology at the University of Illinois. He was in charge of the research laboratories of the Illinois Natural History Survey from 1914 to 1929 and spent his summers from 1914 to 1930 directing marine ecology at the Puget Sound Biological Station. He was also influential in organizing the Ecological Society of America and served as its first president in 1915. Before leaving Chicago, however, Shelford had indoctrinated a number of students with a particular approach to the study of animal ecology that was fostered by the embryological orientation of the department and by the research focus of other departments at Chicago as well. This institutional nexus continued to nurture animal ecology at Chicago in certain directions even after Shelford's departure.

In summary, what were the important organizing themes that helped define the scope and meaning of animal ecology at Chicago in its early years? The strong presence of geography at Chicago under Salisbury and the development of physiographic plant ecology by Cowles certainly drew attention to the importance of understanding the influence of the physical environment on organic life. But the environmental influence idea as developed by Salisbury and other turn-of-the-century geographers was not a theme played out in animal ecology at Chicago. In fact, by the 1920s, the rigid environmental determinism of Chicago geography had yielded to an interactionist paradigm under

Salisbury's student, Harlan H. Barrows, that centered on "man's adustment to the environment, rather than [on] . . . that of environmental influence."[102]

Barrows's later emphasis on human adjustment to the environment points to a more central theme that existed across departmental boundaries and one that had important repercussions for Chicago animal ecology: a focus on the organism-environment relationship as an interactive process. This interactionist model not only pervaded embryological and animal ecology research, but it is also apparent in the pragmatic philosophy of such individuals as John Dewey and George Herbert Mead. Through its behavior and activity, the organism continually restructured its environment to meet new demands and thus had some control over the future course and direction of its development and progress. This idea pervaded ecological thinking at Chicago. It was evident in Davenport's interpretation of adaptation as a "selection of environment fitting a given structure" and in Shelford's continual emphasis on the role of behavior as a determinant of animal distribution. This interactionist perspective also lent support to a physiological view of the organism in its development, because form was interpreted as a consequence of activity. Mendelian genetics found little support in the zoology department at Chicago because it ignored the immediate environmental conditions of the growing embryo, the chemical and physical processes by which the individual characteristics of the organism become expressed. In essence, Mendelian genetics ignored behavior; its focus was, in Child's words, mere descriptive anatomy. Consequently, animal ecology developed at Chicago with little relation to problems in genetics. And the links between ecology and evolution at Chicago came primarily through a Spencerian notion of development implicit in the principle of ecological succession.

Within this framework, animal ecology continued at Chicago despite Shelford's absence. In 1915, Morris M. Wells, a former student of Shelford's, was hired to replace the position left vacant by his adviser. Although Wells continued to teach the ecology course sequence, he did not actively pursue further research and in 1919 he resigned. Wells's replacement, William J. Crozier, himself a researcher in animal behavior, left after one year. The zoology department, already highly inbred, continued its practice of offering vacant positions to former students by recruiting Allee, who was then at Lake Forest College, to the department in 1921. Lillie described the position to Allee as a "succession to the one originally held by Dr. Shelford, and . . . includes more particularly the subjects of Ecology and Animal Behavior."[103]

In the espousal of a developmental view both goal directed and progressive, Chicago biologists had humans as the end stage ever in mind. Their science was not just about the biology of leeches or *Planaria*; it was also about humans, about the place of humans in nature, and the laws of nature that have guided the past and direct the future course of social evolution. A morality could be discerned in nature. Even the succession of plant and animal communities offered prescriptive lessons for human society. This theme of nature as normative has been implied—in Whitman's appeal to specialization and organization as companion principles of progress; in Child's search for a unifying theory of development, heredity, and behavior; in Wallace Craig's discussions of animal life on the North Dakota plains. Yet, this image of biologist as prophet, as someone who could reveal the moral teachings of nature to society through science, has not yet been analyzed. World War I provided opportunities for these social ideals latent within American biology to surface and become expressed in much more explicit fashion. Allee himself returned to the University of Chicago deeply marred by his experiences in World War I and sensitive to the ways in which biology had been brought to bear on moral and social issues raised in the context of the war. His wartime experiences proved pivotal in shaping a program of animal ecology research laden with social meaning.

3

Biology as Gospel

Clyde Allee completed his Ph.D. at the University of Chicago in the spring of 1912. That year tenure-track posts for a doctorate in zoology were scarce, and Allee had to content himself with a one-year position as instructor in botany at the University of Illinois-Urbana. In the fall of 1912, he married Marjorie June Hill, from Carthage, Indiana, a small town about one hundred miles east of Allee's family farm. Clyde and Marjorie shared very similar backgrounds. Both were members of Quaker families with ancestors that had moved from North Carolina to Indiana during the early 1800s; the Quaker communities of Carthage and Bloomingdale, as part of an active region of the Underground Railway, helped the passage of slaves to Canada during the mid-1800s. Earlham College in Richmond, Indiana, was the educational center supported by many of these Quaker settlements. Allee graduated from Earlham in 1908, and Marjorie Hill had entered the college in 1906 but later transferred to the University of Chicago to finish her undergraduate degree in English, which she completed in 1911. Together, the couple ventured on a series of yearly moves from Urbana to Williams College in Massachusetts and then on to the University of Oklahoma in 1914. In 1915, Allee secured a position as professor of biology at Lake Forest College, where the two were able to establish a more permanent home for their two-year-old son.

In the yearly moves from 1912 to 1915, the Allees continually looked forward to at least one constant in their lives: summers at the Marine Biological Laboratory at Woods Hole, Massachusetts. Allee was an instructor in the invertebrate zoology course at the MBL from 1914 to 1918 and the MBL course director from 1918 to 1921 (fig. 3.1). The course provided him with abundant field material for a series of ecological investigations on the factors limiting the distribution of littoral (shoreline) invertebrates in the Woods Hole region. Woods Hole was, however, more than an ideal field station; it was an important institution for establishing a sense of professional identity and

Figure 3.1. The Woods Hole invertebrate class on a picnic on Gay Head, Martha's Vineyard. Warder Clyde Allee is in the white hat and black tie, standing with arms folded. Courtesy of the University of Chicago Archives.

community. Here, biologists, many of whom were isolated during the nine-month academic term in small colleges, gathered to exchange ideas on their latest research and to catch up on changes and happenings of the preceding year. Indeed, "summers spent in research and teaching at the Marine Biological Laboratory in Woods Hole," Allee recalled, "furnished my most important professional experience for the following nine years—much more important than the winters of those years."[1]

Marjorie Hill Allee also looked fondly on those early summers. At Chicago, she studied under the novelist Robert Herrick and gained additional writing experience through her role as women's editor of the student paper, the *Daily Maroon*. In her years at Lake Forest, she contributed reviews to the book page of the *Chicago Daily Tribune* and began writing for the *Youth's Companion*. This launched her into a career as a professional writer of young women's novels, and she often used her acquaintances with the professional biological community at Chicago and Woods Hole as background material for her books (fig.

Figure 3.2. Marjorie Hill Allee at age 19. © The Horn Book, Inc. Reprinted by permission of the Horn Book, Boston.

3.2). *Jane's Island* is one such novel, written in 1931 to capture the family's summers at the MBL "before brick buildings and expensive apparatus and summer people had got the better of the simple Woods Hole." Like many of Marjorie Hill Allee's books, *Jane's Island* is peppered with moral lessons and offers a unique glimpse into the events that shaped her husband's early professional career.[2]

Jane, the namesake of *Jane's Island,* is a spunky child, an expert in *Planaria* collecting, and the daughter of a Dr. Thomas, who has brought his family to Woods Hole for the summer while he pursues scientific research at the MBL. The Thomases have hired Ellen McNeill, a young woman enrolled at the University of Chicago, to be Jane's "guide, philosopher, and friend," and the story details the different encounters faced by Jane and Ellen in coming to grips with the adult world. The companions quickly find themselves enmeshed in a rivalry between Dr. Thomas and a Dr. von Bergen, a German scientist who loosely resembles the real life biologist Jacques Loeb, befriended by

Thomas in their graduate days. Since the war, however, von Bergen has become disillusioned, acrid, and pessimistic. When Thomas asks von Bergen to join him in singing German songs from their youth, von Bergen bitterly remarks, "Zose days are past. Zose days when we believed in brothership among all living sings and many ozzer foolish dreams."[3]

Von Bergen has come to try and discredit Thomas's scientific work. Thomas has spent the last ten years trying to understand why animals form groups. His experiments seemed to demonstrate that group life provides benefits to the individual in its struggle with the environment. But since the war, von Bergen considered Thomas to be "a sentimental soul" who got "the results" he wanted "to get; that [Thomas] likes to think of the animal kingdom living in cooperation, whereas [von Bergen] believes the truth of it is that every animal lives at the expense of others." The theme of cooperation also filters into the practice of scientific research. Thomas insists that his children help von Bergen find the experimental animals necessary for his research, much to their dismay. And the need for research funds for Thomas's work prompts Mrs. Thomas to hold an afternoon tea to which Mrs. Smith, the wife of a wealthy philanthropist, is invited. But Jane in her childish innocence has little political savvy and offends Mrs. Smith, seriously jeopardizing Thomas's chance at getting any foundation support. In the end, however, the true scientific spirit wins out over egos, avarice, and the search for fame. When Thomas is rushed to the hospital for an appendectomy, von Bergen breaks into Thomas's lab and saves his experiments from ruin, because his own researches have convinced him that Thomas's theories are indeed correct. Cooperation and community, the quintessential elements of science, are achieved when, in true altruistic fashion, von Bergen recommends to the foundation that his own grant be transferred to support Thomas's work.[4]

Jane's Island, although fiction, captures elements of Allee's life with a naïveté that strips them of the complexities and equivocations that mark historical experience. Dr. Thomas is the idealized character of Allee. War, cooperation, and the quest for funds and acceptance are the dramatic motifs into which both characters are cast. Immediately after the First World War, Allee initiated a research program on the causes and significance of animal aggregations, on the "physiological effects of crowding upon the individuals composing the crowd."[5] Allee hoped this research would bridge the widening gap between autecology and community studies, yet it also served a political agenda. His experiences as a pacifist during the First World War brought him in contact with social reformers and liberals of Chicago who were strongly identified with the political left. After the war, Allee propelled himself into "undertaking scientific experiments designed to throw light" on the subject of war, and he incorporated the social message of the liberal pacifist into his work on animal aggregations.[6] Ellen's musings in *Jane's Island* reveal this message in its most simple form. "If even those tiny brown bits of planarians are better off for living together," Ellen reflected, "perhaps we can be more certain that . . . quarrels and wars aren't exactly natural things to expect." "Bless you!" exclaimed Miss Wareham, a character most likely based on the biologist Cornelia Clapp, a Whitman Ph.D. student who taught zoology at Mt. Holyoke. "That is poetry! No scientist would jump that far at one leap." Perhaps, but only in the fictionalized world of science.[7]

The Politics of Pacifism

The profound impact of the war on Allee was rooted in his strong religious upbringing. Allee was born on June 5, 1885, a birthright member of the Society of Friends. His father, John Wesley Allee, was the son of a Methodist minister, and became a "convinced Friend" on his marriage to Mary Emily Newlin, whose ancestors had established the Quaker settlement of Bloomingdale. Despite his membership in the Quaker church, John Allee did not completely renounce his Methodist affiliations. He often ushered the family to revival meetings at the Methodist or United Brethern churches in the area. Allee recounted with anguish his own conversion experience at a revival meeting as a child of eleven. Feeling the pressure to conform and fearful of a life condemned to hell, Allee told his parents that he was converted, although secretly he felt no different. At college, he "conformed to the strong religious atmosphere" of Earlham, despite a "growing honesty within himself" about his own religious convictions.[8]

Allee's religious beliefs were further shaken on entering graduate school at the University of Chicago. He had enrolled in a course on animal evolution taught by William Lawrence Tower, a controversial figure who resigned from the department in 1917 in the wake of a highly publicized divorce trial.[9] Allee "was hardly prepared for" Tower's opening sentence, "which was: The theory and teaching that there is a God is a lie." "I felt sorry for that misguided man," wrote Allee. "I had often heard of such people—infidels, Atheists and that sort but he was the first I had met and I planned to show him the error of his ways, but he was called away suddenly to see to an experiment in Central America and when he returned I was not ready to talk with him because facts had been coming home that were making me think in a different way from ever before." Allee quickly learned that many of the natural phenomena that he explained since his childhood as the "working of the Hand of God" were solely "explicable in physical-chemical terms."[10] By the time he received his doctorate in 1912, he looked skeptically on vitalists that left a "large place in their thinking to be filled in by supernatural means." He had come to embrace a mechanistic approach to the study of life, and often expressed his philosophical stance in the following terms. The "mass of unknown facts," Allee speculated, could be represented "mathematically as *X* without trying to postulate what *X* may be. If you think as Loeb does that the unknown is relatively small," he continued, "make a tiny *x* but if it appears immensely large go out on the sand and draw an *X* commensurate with your idea but by all means keep the idea that this is an unknown and not an unknowable quantity and that the mechanistic view is the only working hypothesis of use in solving the problem." From a methodological standpoint, Allee proceeded under the assumption that physicochemical explanations could in principle account for all natural phenomena. He eschewed attempts to sneak vitalistic explanations in through the back door, and looked upon the physiological investigations of Child and Loeb as models for biological research.[11]

Allee advocated a mechanistic position in the study of biology, yet he did not reject the religious doctrines of his Quaker upbringing when contemplating human affairs. For the general public, pacifism and nonviolence are perhaps the most outstanding precepts of the Quaker faith. Although the practice of conscientious objection stems from the Quaker Declaration of 1661, which renounced "all outward wars and strife and fightings with outward weapons for any end or under any pretense whatever," pacifism is still regarded as a personal choice among individual Quakers. For Allee, however, adherence to pacifism, understood in the strict renunciation of any war, remained a personal conviction throughout his life.[12]

Prior to World War I, the American peace movement was identified with such respected organizations as the Carnegie Endowment for International Peace and the World Peace Foundation. Membership in these organizations came largely from conservative or moderate factions that equated peace with economic order and stability. In the prewar years, the movement was largely legalistic and educational and possessed a "patriarchical, and elitist quality" that deprecated radical political action.[13]

When the United States entered the war in April 1917, the conservative element of the American peace movement abandoned its previous idealism and supported Wilson's call for a "war to end war." Those committed to pacifism under any circumstances found themselves increasingly alienated and subjected to public persecution. As Charles Chatfield has argued, "The word pacifist changed under the pressure for patriotic conformity in 1917–1918. Having had the benign connotation of one who advocated international cooperation for peace, it narrowed to mean one who would not support even a 'war to end war.' Pacifists were linked with draft dodgers, socialists, and communists, portrayed in hues from yellow to red."[14]

Allee's commitment to pacifism brought him into the arena of this emerging movement. In March 1917, he was appointed chairman of the Quaker War Service for civilian relief in Chicago, which formed as an outgrowth of the Chicago Monthly Meeting of Friends. Initially, the activities of this committee focused on conscription and the mistreatment of conscientious objectors (COs) in military camps and on opposition to mandatory military training in schools. In December 1916, for example, the committee organized a protest against the abuse of conscientious objectors in England.[15] Upon America's entry in the war, however, the threat of conscription became even more immediate. On May 18, 1917, a provision was added to the Selective Service Act that enabled religious conscientious objectors to partake in noncombatant service. The Society of Friends hoped that the American Friends Service Committee (AFSC), established in April 1917 to aid relief and reconstruction work abroad, would qualify as service for the COs. This hope was not realized. Throughout the summer and winter of 1917, COs were forced to attend military training camps. Many suffered severe beatings and were placed in solitary confinement for their refusal to take part in military exercises. Conscription became a focal point of concern for such organizations as the AFSC, the Civil Liberties Bureau, and the Woman's Peace Party. It was in this context that Allee endorsed a pamphlet entitled "What Happens in Military Prisons: The Public Is Entitled to the Facts," distributed by the Chicago branch of the Civil Liberties Bureau in December 1918.[16]

The Quaker War Service for civilian relief became the Chicago regional committee of the AFSC, and Allee served as either chairman or vice-chairman of the regional committee from 1917 to 1950. His associations with the AFSC brought him in contact with other peace organizations and social liberals of Chicago such as Jane Addams, who played an influential role in establishing the Woman's Peace Party, an organization that became the Women's International League for Peace and Freedom. At the close of the war, the Allees worked closely with Addams trying to secure help from the German societies in the city to aid in Central European relief, and Allee lectured occasionally to the Hull House summer school in the 1920s.[17]

Through these associations, Allee became exposed to a peace movement that was strongly identified with the political left.[18] For the liberal pacifist, war itself was "an integral part of an unjust social order," which, by means of violence and authoritarianism, ensured either a conservative form of social control or revolutionary change.[19] Rejecting a politics of domination, the pacifist attempted instead to "fashion a new conception of peace as a political exercise in the nonviolent pursuit of change."[20] Peace and cooperation were parts of this political process. Democracy was not to be understood in terms of power amassed by conservative groups to ensure social order; rather, political power should be conferred on members of society at large.

The argument for a distributive democractic process reflected a philosophical position at the heart of liberal pacifism: the importance of individual worth. The individualism of liberal pacifists, however, differed significantly from the egoistic Spencerian notions of individualism and laissez-faire. The individual did not live and profit at the expense of all others. Instead, both the individual and the state were responsible for preserving individual welfare and rights. Society was ordered for the benefit of its members. Conscription was thus a direct violation of this principle, because it took away the personal rights of the individual which the state was supposed to protect.

Although many liberal pacifists were members of the Socialist party, their distrust of authoritarianism could lead them to embrace an almost conservative, libertarian political stance. Such was the case with Allee. While Marjorie Allee supported Norman Thomas and the Socialist party, Allee often backed Republican candidates, depending on their military stance. He feared the concentration of governmental authority, denouncing, for instance, the New Deal of Roosevelt because of its close alliance with big business and its attempts to expand the power of the executive branch. Although sympathetic to socialist causes, Allee never became a member of the Socialist party because of his extreme antiauthoritarianism and the value he placed on individual rights.[21] This position was not necessarily uncommon or contradictory. Norman Thomas, an active pacifist during World War I and presidential candidate for the Socialist party from 1928 to 1948, characterized this dilemma before joining the Socialist party in 1918. "The ultimate values in the world," Thomas wrote, "are those of personality, and no theory of the state, whether socialistic or capitalistic, is valid, which makes it master, not servant of man."[22] Elsewhere, Thomas argued, the individual is a "product of the group, but the group is only valuable as it permits personalities, not automatons to emerge."[23]

By 1917, a pacifist in America assuredly faced public ridicule. In *Peace and Bread,* Jane Addams described her own experiences as a pacifist during World War I. "After the United States entered the war," she wrote, "the press throughout the country systematically undertook to misrepresent and malign pacifists as a recognized part of propaganda and as a patriotic duty."[24] Allee did not escape this propaganda campaign. With Germany's renewal of unrestricted submarine warfare at the end of January 1917, relations between Germany and the United States became more and more tense. President Wilson's breaking of diplomatic relations with Germany in early February sparked a heated call to arms in newspapers across the country. In Chicago, the *Tribune* published editorials daily that favored universal military training and military preparedness. Local Chicago organizations that advocated peace met on February 7 to urge Wilson to call a conference of neutral nations and to subject any Congressional declaration of war by the United States to a national referendum.[25] In the midst of these heightened tensions, Allee delivered a chapel talk at Lake Forest College on conscientious objectors to military service. The lecture engendered a number of negative editorials in local Illinois papers. The *Springfield News-Record,* reporting the story, captured local sentiment when it referred to Allee's conscientious objection as a "most convenient theory." "Sometimes war is unavoidable," the writer commented, "and college professors are no more necessary to civilization than carpenters and cobblers. In fact, if we had to dispense with one or the other, we should prefer to give up the professors."[26] The college administration reprimanded Allee by forbidding him to give further chapel talks. But the administration could not so easily silence its faculty. Allee continued to speak out on the war. He had his "salary docked for telling a class that he personally did not believe in war" and received "considerable incorrect newspaper comment for stating that he preferred giving what little money he could spare to war relief rather than war bonds."[27]

Two months later, after the scandal over Allee's talk on conscientious objectors had blown over, another controversy ignited. On April 7, one day after Wilson's declaration of war, George W. Schmidt, chair of the German department at Lake Forest was arrested for alleged slurs against the U.S. government that he made in front of a crowd while vacationing on his family farm in Stanley, Wisconsin. Schmidt was "absent on leave" for the 1917–18 academic year and was officially dismissed in May 1918. President Nollen, trying to uphold the college's conservative image, declared immediately after Schmidt's arrest that "the faculty were unanimous in their patriotism, if not in their opinions concerning the war," despite Allee's continued vocal opposition.[28] The Schmidt incident is important because it touched the lives of two other individuals who were to become Allee's closest colleagues and friends:

Alfred E. Emerson and Karl P. Schmidt. Allee, Emerson, and Schmidt were deeply influenced by the war, each in his own way, and this influence reverberated throughout their scientific careers.

Karl, the son of George Schmidt, was born in Lake Forest in 1890 and enrolled in Lake Forest College as a freshman in 1906. He left the college after one year and spent the next six years managing his mother's dairy farm in Stanley, Wisconsin (where his father would be arrested). Farm life proved to be a source of inspiration for Schmidt's interest in natural history, and in 1913, at the urging of his Lake Forest teacher, James G. Needham, Schmidt enrolled as a sophmore at Cornell University where Needham had assumed a position as professor in entomology and limnology. While at Cornell, Schmidt lived in the Needham home and helped with the family's Lake Ontario farm, which Needham often used as a field site for his limnology classes. He also assisted in Needham's course on the natural history of the farm.[29] The mix of agricultural life and natural history was a common undercurrent in the life sciences at Cornell. The site of an agricultural experimental station, Cornell was also home to Liberty Hyde Bailey and Anna Comstock, both active in Theodore Roosevelt's Commission on Country Life which sought to instill the values of rural life into the minds of young children through nature study.

Schmidt planned to return to the family farm in Wisconsin in the summer of 1916 after accompanying the Cornell Geological Expedition to Santo Domingo, but on his return he met Mary Dickerson, curator of the Department of Herpetology at the American Museum of Natural History, and she offered him a position unpacking reptile collections from the American Congo Expedition. The war interrupted Schmidt's position at the museum. On his father's arrest, he wished to return to the family home in Stanley, much to the dismay of his mentor Needham.[30] He was drafted in the fall of 1917, but never saw active combat. After the war, he resumed his association with the American Museum and became assistant curator of reptiles and amphibians. In 1922, he accepted a position as curator of the Division of Reptiles and Amphibians at the Field Museum of Natural History in Chicago, and shortly thereafter his lifelong association and friendship with Allee began.

At Cornell, Schmidt had developed a very close relationship with Alfred E. Emerson and inspired in Emerson a keen interest in natural history. Emerson was six years younger than Schmidt and had grown up in a family environment of academic life accented by a parental reverence for the arts and humanities. His early childhood was spent in the vicinity of Cornell, where his father was a professor of classical archaeology. The family moved to Chicago in 1905 when his father became curator of antiquities at the Art Institute of Chicago, but Emerson returned to Ithaca in 1914 to pursue an undergraduate education at Cornell. He taught nature study classes for Anna Comstock and graduated from the Department of Entomology in 1918.[31]

The correspondence between Emerson and Schmidt during their Cornell years is colored by their wartime fears and experiences. Emerson was disheartened by the increased anti-German sentiment in his country but in his naïveté refused to believe that such mob attacks as experienced by Schmidt's father could occur anywhere outside a "narrow-minded country village" like Stanley, Wisconsin. He found it difficult to reconcile his own support of the U.S. entry in the war with the suffering and sadness that had overcome his German friend. "I really believe this war is a war against war," Emerson confided in Schmidt, "and therefore I don't suffer under it as you must but look at its success and outcome as a necessary advance in human progress." When Schmidt was drafted, Emerson could not understand how he "could go into the thing with no heart." In contrast, Emerson enlisted in the summer of 1918 with patriotic fervor.[32]

World War I struck the lives of Allee, Emerson, and Schmidt, as it did, to a greater or lesser degree, other biologists in America. Its impact depended, to a large extent, on personal belief and individual circumstance. Of the three, Allee was perhaps the most affected. His firm commitment to pacifism continually placed him at risk of jeopardizing his professional career. Schmidt's own personal encounters with an anti-German public, along with Allee's later friendship, may have led him to join the Chicago 57th Street Meeting of Friends and take an active interest in the activities of the AFSC during the Second World War. Emerson occupied a more mainstream position during World War I, and although the war touched his life, it did so in a tangential way. The circumstances that united these individuals' personal lives, also influenced their biological vision. During the war, a substantial literature had surfaced in the American biological community denouncing war on biological grounds. Allee was well acquainted with these writings, and from them crafted a research program within the domain of ecology to which Emerson and Schmidt would later subscribe, a program centered on cooperation as a fundamental principle of community integration and a dominant force behind ecological and evolutionary change.

The Biology of War

While the major battles of World War I were being fought in the trenches abroad, the tremors of war deeply shook the thoughts and feelings of many biologists in America. The sinking of the *Lusitania* on May 7, 1915, seriously challenged America's own neutrality and raised questions among American scientists about their own possible contributions to the national welfare in the event of war. In June 1916, with the establishment of the National Research Council by the National Academy of Sciences, American scientists became actively involved in the growing war effort.[33] Overshadowed by the flurry of war-related research in the physical sciences, biologists struggled to find their own niche in the science-government sector. At their annual meeting in 1917, the American Society of Zoologists, to which Allee served as secretary from 1917 to 1921, dedicated an entire session, "The Value and Service of Zoological Science," to questions on this topic. Similarly, biologists discussed the contributions of zoology to human welfare at the Pittsburgh meeting of the zoological section of the American Association for the Advancement of Science (AAAS) during the same year. But for many biologists, the impact of zoology extended beyond the practical applications emphasized at these symposia. Maurice Bigelow of Columbia University reminded AAAS members in his address that "we have overlooked the fact that a philosophical application of a pure-science theory may come to be a guiding force in the material affairs to which science is directly applied. Such is the case in the relation of certain phases of evolutionary philosophy to the Great War."[34]

The precise evolutionary philosophy Bigelow had in mind was the Darwinian emphasis on the survival of the fittest in a nature dominated by ruthless struggle. "Pure-science theory," in this case Darwinism, had, according to Bigelow, guided and directed the course of German militarism and its philosophy of the superior state. How much Darwinism actually contributed to the formation of the Pan-German League and, hence, to the war is a question open to dispute.[35] For a number of American biologists, however, there was no question; in numerous presidential addresses and popular articles, prominent American biologists such as Leon J. Cole, Vernon L. Kellogg, David Starr Jordan, William Patten, Raymond Pearl, and William Ritter attacked what they perceived as a distorted view of Darwinian evolution taken by Germany as justification for international war.[36]

For American biologists, Vernon L. Kellogg's *Headquarters Nights,* written as an exposé of the horrors of German ideology and its underlying evolutionary philosophy, provided compelling evidence that evolutionary theory was at least partly responsible for the war.[37] Kellogg, an entomologist by training, was a prolific writer and popularizer of biological subjects, especially evolutionary topics. As professor of entomology at Stanford, he became an intimate friend of

David Starr Jordan, an ichthyologist and ardent peace activist who served as Stanford's first president. Kellogg and Jordan collaborated on many projects; they taught a highly successful course on evolution and coauthored a number of biology texts. The shared interests of Kellogg and Jordan also extended into political affairs.[38]

As the fighting in Europe continued, Kellogg was drawn into the political arena. He was very active in the formation of the National Research Council (NRC) and became chairman of its divisions of agriculture, botany and zoology. With the permanent establishment of the National Research Council in 1920, Kellogg was named permanent secretary, a position he held until his death. Kellogg's humanitarian and pacifist convictions led him to relief work abroad. In 1915, he joined the staff of the Commission for Relief in Belgium (CRB) and Northern France. *Headquarters Nights* is an account of Kellogg's experiences as the CRB's chief representative and his associations with the German higher command. But *Headquarters Nights* is more than a documentary. As Kellogg himself describes it, "*Headquarters Nights*" is the confession "of a converted pacifist."[39]

The headquarters of the German General Staff, situated on the Meuse in northern France, was also Kellogg's place of residence for several months while serving as representative of the American Relief Commission during 1915. Living among German officers, Kellogg had many opportunities to discuss the weltanschauung of the German people. Through conversations with a Professor von Flussen (Kellogg used a fictitious name to hide the person's real identity), a professor of zoology and second commandant of the headquarters town, Kellogg leads the reader to the inevitable conclusion that the German point of view is based on "a whole-hearted acceptance of the worst of Neo-Darwinism, the *Allmacht* of natural selection applied rigorously to human life and society and *Kultur*."[40] In the preface to *Headquarters Nights,* Theodore Roosevelt wrote: "The man who reads Kellogg's sketch and yet fails to see why we are at war, and why we must accept no peace save that of overwhelming victory, is neither a good American nor a true lover of mankind."[41]

At the root of the evil, Kellogg saw a "Germanized" Darwinism, with its undue emphasis on natural selection and the struggle for existence. "Professor von Flussen," Kellogg remarked, "is Neo-Darwinian, as are most German biologists and natural philosophers. The creed of the *Allmacht* of a natural selection based on violent and fatal competitive struggle is the gospel of the German intellectuals; all else is illusion and anathema."[42] From this "distorted" biological philosophy, the justification of war naturally followed:

> That human group which is in the most advanced evolutionary stage as regards international organization and form of social relationship is best, and should, for the sake of the species, be preserved at the expense of the less advanced, the less effective. It should win in the struggle for existence, and this struggle should occur precisely that the various types may be tested, and the best not only preserved, but put in position to impose its kind of social organization—its *Kultur*—on the others, or alternatively, to destroy and replace them.[43]

Faced with such arguments, Kellogg was forced to abandon his pacifist beliefs; the only solution was to fight the war to its end.

Philip Pauly has suggested that the war "brought home the difficult cultural position of biology" because, unlike their colleagues in physics, chemistry, or even psychology, biologists contributed little to the practical side of the war effort. *Headquarters Nights* was, however, the perfect propaganda piece, not only for the American government interested in arousing anti-German sentiment but for biologists as well. Indeed, Alfred Kelly's claim that most German militarists used the rhetoric of struggle without being Darwinists suggests that the supposed connection between Darwinism and German militarism was created by Kellogg and others for propaganda purposes.[44]

The war created a unique opportunity for American biologists. By associating the doctrine of natural selection with German military ideology, these biologists made the issue more than a debate about science, but one about culture as well. Drawing upon evolutionary theory and their own moral convictions, they were able to condemn Germany's militarist philosophy on biological grounds, thereby contributing to the American war effort while reinforcing the profession's own cultural worth.[45] Although there were many variants of antiwar biology, three major arguments were advanced by American biologists during the First World War. The first was to deny the *Allmacht* of natural selection in evolution, an approach consistent with many biologists' views of evolution during the period. The second line of attack bolstered support from the popularity of eugenics and focused on the dysgenic or ill effects of war that ultimately led to racial deterioration. Both Kellogg and Jordan were the champions of this cause. The third position had its roots in the social evolutionist literature of the late 1800s, in such writers as the French sociologist Alfred Espinas and the Russian Prince Peter Kropotkin, who emphasized the value of mutual aid in the organism's struggle with its environment and in the development of social life. It was this particular strand of antiwar biology, especially with its links to the writings of Spencer and Espinas, that Allee would develop into a program of ecological research. None of the three positions, however, rejected the biological determinism at the heart of German war ideology. Indeed, no American biologist questioned the legitimacy of German war ideology by arguing that evolutionary theory was not, in principle, applicable to human affairs. To have done so, would have seriously undermined the evolutionary biologist's authority within American society, especially in the domain of morality and social affairs.

A Eugenics of Peace

In *Headquarters Nights,* Kellogg accused Germany of wholeheartedly accepting August Weismann's claims for the *Allmacht,* or all-sufficiency, of natural selection. During the latter part of the nineteenth century, Weismann, a renowned cytologist and professor at Freiburg, became a staunch defender of the all-sufficiency of natural selection in accounting for evolutionary change. Weismann postulated a theory of heredity in which the germ plasm was completely separated from, and unaffected by, the somatic (or body) cells, a position obviously at odds with the precepts of Lamarckism.[46] Kellogg looked with disfavor on Weismann's dogmatic adherence to natural selection, and he held Weismann and his followers responsible for the "distorted" views of Darwinism prevalent in Germany.

Weismann's scientific dogmatism opened the door for attack. During the first two decades of the twentieth century, biologists seriously questioned the importance of natural selection as a species-forming agent. Many of the characteristics that separated one species from another were regarded by naturalists as nonadaptive, and, therefore unexplainable by natural selection alone. Other mechanisms such as geographical isolation, favored by field naturalists such as David Starr Jordan, offered alternative explanations to account for the origination of species. Experimental biologists had their own preferred mechanisms for species formation: DeVries's mutation theory and Mendelism were two of the most likely alternatives.[47] Since many American biologists regarded natural selection as one of many possible evolutionary mechanisms, rigid adherence to natural selection as *the* causal agent of evolutionary change was suspect. Hence, biologists could dismiss Germany's biological justification for war because it was founded, according to Kellogg, on the all-sufficiency of natural selection, a position held in biological contempt. Raymond Pearl, a mathematical population biologist at Johns Hopkins, criticized German biological philosophy on precisely these grounds: "Nowhere in nature does natural selection, as indicated by modern careful study of the subject, operate with anything like that mechanistic precision which the German political philosophy postulates. . . . Nature often does not operate on the natural selection basis, though logically—at least in formal logic—it ought to."[48] Pearl thus exposed one apparent biological flaw in Germany's justification of war.

Selection itself, however, could be used to denounce the biological philosophy of war. Few biologists were more committed to exposing the horrid effects of war than Jordan. A leading peace activist, Jordan served as vice-president of the Anti-Imperialist League in 1898, chief director of the World Peace Foundation from 1909 to 1914, and chairman of the Emergency Peace Federation in 1917. Unlike Allee, however, Jordan was a member of prewar peace organizations, moderate in their stance, that rallied behind Wilson's declaration of war. "Our country is now at war and the only way out is forward," Jordan told the *San Francisco Bulletin* in a statement that mirrored the rhetoric of American conservatism. "We must now stand together," Jordan asserted, "in the hope that our entrance into Europe may in some way advance the cause of Democracy and hasten the coming of lasting peace."[49] Kellogg also joined his colleague's cause and renounced his pacifist convictions. With eugenics as their most popular biological ally, Kellogg and Jordan stripped German militarism of its Darwinist foundations.[50]

Although the militarists argued that war led to racial betterment by eliminating the weak, unfit portions of humanity and selecting out the strong, Jordan and Kellogg saw otherwise. In numerous articles and addresses they cited evidence to indicate that war led to the promotion of ill birth, or dysgenics, rather than character improvement. Military selection was not analogous to natural selection because it was ultimately maladaptive.[51] The military, these peace eugenists pointed out, is not a cross-section of the population but, instead, consists of those individuals possessing superior characteristics such as "ripe youth, full stature and strength, and freedom from infirmity and disease."[52] Because these individuals fight while the infirm do not, a higher number of individuals of genetic worth are eliminated while the less desirable members of the population produce a greater portion of offspring at home. War thus led not to race improvement but to racial deterioration. As part of a study for the Carnegie Endowment for Peace, Kellogg conducted a statistical analysis that demonstrated the Napoleonic Wars had decreased the average height of the French male.[53] But the consequences of war went much farther than the lowering of stature. Eugenists cited numerous instances of racial deterioration resulting from warfare, including the downfall of Rome, with its obvious analogy to the war situation in Europe.[54] Thus, even within the confines of Darwinism, a case could be made against the biological philosophy of war. War did not preserve the fit, but the unfit, ultimately leading to racial extinction. Eugenists like Jordan and Kellogg transformed the German military precepts of Darwinism into a doctrine for peace.

Cooperation, Community, and Progress

A third antiwar biology strand focused on the meaning of struggle and the "survival of the fittest." American biologists accused German militarists of equating fitness with physical power whereby the strongest and fiercest eliminated the weak. Strength, however, need not mean pure physical might. Sociality was itself a powerful force in combating environmental struggle. As Jordan remarked, "altruistic social adjustments are powerful factors in the struggle for existence in the life of animals as well as man."[55]

Arguments for mutual aid and cooperation had their roots, ironically, in the social evolutionist writings of Herbert Spencer. Spencer saw evolution as the developmental progress from the simple to the complex, from the homogeneous to the heterogeneous whereby increasing specialization and integration marked the higher stages of both individual and social life. In advancing the analogy between the individual and social organism, Spencer embraced a functionalist perspective, embedded in physiological metaphors of the organism and the body politic, that emphasized the interdependence and integration of parts into a unified whole. "All kinds of creatures are alike," Spencer wrote, "in so far as each exhibits co-operation among its components for the benefit of the whole; and this trait, common to them, is a trait common also to societies."[56] Adaptation of the part, be it a cell or an individual, to the whole thus resulted in the mutual dependence of living forms on one another. Although one commonly identifies Spencer with a Victorian ideology of industrial capitalism centered on cutthroat competition and laissez-faire individualism, such was not the case in his early writings. Robert Richards has persuasively shown how Spencer's idolatry of the individual brought him, in his younger years, to envision a utopian socialist state where "social distinctions of class and caste would be absorbed in a respect for the dignity of each person's share in the division of labor."[57] By the 1870s, when Spencer published the first volume of *Principles of Sociology,* such ideals were abandoned in favor of conflict and struggle as the driving force behind evolutionary progress, ideas for which he is perhaps best remembered. Yet other intellectuals returned to elements of Spencer's early optimistic utopia in their own cooperationist writings.

During the late nineteenth century, a number of individuals took issue with the image of nature, red in tooth and claw, portrayed in the writings of Darwin, Thomas Henry Huxley, and a mature Spencer. One need only recall that Darwin had used the "term Struggle for Existence in a large and metaphorical sense, including dependence of one being on another" to invoke a more benign vision of nature in which cooperation and association were the driving forces behind evolutionary change.[58] Even Darwin, although a thoroughgoing Malthusian, traced the evolution of the moral sentiments to the "demands of the social instincts" in *The Descent of Man*.[59] But Darwin's primary focus on biotic struggle drew attention away from the interaction between individuals and their abiotic environment, which was the subject of early animal ecology research. The emphasis placed on abiotic struggle, development, and progress in the writings of the French sociologist Alfred Espinas and, to a lesser extent, Kropotkin were of far more importance than Darwin to Allee as he later outlined his own arguments for a theory of sociality that blended ecological science with his own political experiences during the First World War.

In *Des Sociétés Animales,* Espinas merged the cooperationist and individualistic themes that marked Spencer's youth. He favored Spencer's emphasis on the continuity between organizational levels, between the individual and society, over the views of his fellow countryman Auguste Comte, who stressed the emergence of society as a discrete and distinctive stage in human progress. Association, Espinas argued, was a powerful developmental force throughout the animal kingdom. "Many of the highest social animals," Espinas exclaimed, "conduct themselves as if each individual member of the group was of absolute value for the others. . . . Far from the struggle for existence, far from the oppression of individuals being the characteristic trait of life within the confines of the individual and of society, it is the coalition for better survival in this struggle, it is the respect of the individual which is the first condition and dominant characteristic."[60] Interdependence and association were thus essential features in the organism's struggle with its environment. Like Spencer, Espinas adhered to a functionalist view of society in which the individual members were adapted to meet the needs and conditions of the growing social organism. Yet, in an interesting turn of phrase, Espinas declared that not only was the society an organism, but the organism was itself a society. Hence, sociality was coextensive with life itself.

Individualism and organicism are intertwined in *Des Sociétés*

Animales. Espinas upholds the priority of individual rights; yet the individual is itself subsumed under the social organism to which it, along with its fellow members, gave birth. The difficulty is in balancing the rights of the individual, so dear to industrial capitalism, with the demands and needs of the state, especially as nationalistic fervors began to ferment through Europe during the latter part of the nineteenth century. Espinas's explanation of the collective conscious reflected these growing concerns with *l'amour de patrie*. "One can affirm as a general law," he noted, "that the distinctness with which social consciousness arises is in direct relation to the strength of its hate for a stranger. Altruism is then truly an extended self, and social conscious is an individual conscious."[61] Nationalism enhanced organic solidarity, spawning the true social self. Interestingly, Allee, who revered Espinas's work, ignored the nationalistic assumptions implicit in *Des Sociétés Animales*.

While Espinas blended individualism and nationalism under the guise of organic solidarity, Kropotkin looked to the more primitive animal associations as a model for political life. Kropotkin, a leader of the international anarchist movement, settled in London in 1886. During the years 1862–66, he led a series of expeditions through unexplored regions of Siberia, and his zoological observations later provided important biological evidence for his work, *Mutual Aid*. Published first in serial form by the *Nineteenth Century* between 1890 and 1896 and in book form in 1902 and 1914, *Mutual Aid* was, in some respects, an attempt to resolve the conflict between evolution and ethics that Huxley presented in a series of essays culminating in his Romanes lecture at Oxford in 1893. Writing in direct response to Huxley's portrayal of nature as a "gladiator's show," Kropotkin set out to undermine the "distorted" Darwinian views that portrayed all of nature as struggle and the worldview that ensued. The undue emphasis placed on struggle between individuals, instead of the real struggle between the organism and its environment, resulted in a false portrayal of nature. The fittest, Kropotkin argued, are not the strongest but are those animals that have acquired the habits of mutual aid.[62]

Darwin and his followers had, according to Kropotkin, attributed far too much importance to biotic struggle. Like many of his Russian colleagues, he rejected the Malthusian argument at the heart of Darwin's evolutionary theory. Competition was never as severe as Darwin proposed, because environmental factors such as climate reduced the level of the population before competition became a significant factor. His observations of animals in Siberia suggested that the abiotic environment was a more important influence in the organism's life than biotic struggle. Hence, Kropotkin directed attention away from competition toward organismic interactions that alleviated its severity. Traits that furnished a release from competition were selected for. Cooperation arose through the process of natural selection because it was adaptive, not in the Malthusian struggle between individuals of the same species but in the organism's struggle with the physical environment. Through mutual aid and sociality, the organism was better able to ameliorate the harsh conditions of its physical surroundings.

Neither Espinas's nor Kropotkin's theory dealt with the main problem of Darwinian evolution; namely, descent with modification. Neither theory, in and of itself, could explain the appearance of individual variation or the transformation of species. For this, both Kropotkin and Espinas resorted to modified versions of Lamarckism.[63] Their evolutionary views were rooted in a progressive developmental metaphor gleaned from Spencer that accounted for the increasing interdependence and association of organic life. It was a perspective that both informed and meshed with ecological theory in the early twentieth century.

These cooperationist renditions of Spencerian evolutionism tied to an ecological emphasis on interdependence received further support during the First World War. In 1914, *Mutual Aid* was reprinted at the request of a London *Times* reader, who denounced the use of Darwinian theory as an explanation for the war with Germany and called for a work that emphasized the importance of cooperation. P. Chalmers Mitchell's *Evolution and the War,* published in 1915, was another British work in the genre of antiwar literature. Mitchell, trained in natural science at Oxford and secretary of the Zoological Society of London from 1902 to 1935, was one of the few biologists to question the deterministic assumptions underlying both war and antiwar biology. Although its antideterminism was uncharacteristic of antiwar biology, *Evolution and the War* is illustrative of the way in which ecology supported this cooperationist perspective.[64] Mitchell spoke favorably of Victor Shelford's *Animal Communities in Temperate America* for depicting a more benign struggle in nature, where interdependence and unconscious cooperation facilitated adaptation to the environment. Somehow Mitchell glossed over the first chapter in Shelford's book. Using a lengthy passage from Theodore Roosevelt to depict the violence of nature, Shelford ended the section with a reminder that "to kill is nature's first law."[65] But the study of communities and their integration seemed to Mitchell to refute the interpretation of "Darwin's 'metaphorical phrase,' the struggle for existence in any sense that would make it a justification for war."[66]

Within the United States, these various renditions of Spencerian social evolutionism took on special significance during the First World War as a battle over evolution's true meaning gained momentum within the nation's boundaries. While Kellogg's *Headquarters Nights* appealed to members of the American biological community, it was also used by antievolutionists such as William Jennings Bryan as evidence for the dismal side of Darwinism.[67] World War I was a crucial turning point for the fundamentalist movement in America. The prewar struggles between fundamentalism and modernism were largely confined to disputes over interpretation of theological doctrine. Advocating a literal reading of the Bible and adopting a premillennial view of history, conservative evangelicals largely refrained from political or social debate. Rejecting the social reform elements of liberalism, salvation was seen as individual and private, dependent on one's trust in the redemption of Christ. The war, however, breathed a social and political life into the fundamentalist movement, as modernism became associated with German Kultur, threatening to undermine the whole moral fabric of civilization.[68] Citing *Headquarters Nights,* fundamentalists like William Jennings Bryan implicated the doctrine of evolution as the breeding ground of German "barbarism." Had not Kellogg himself pointed out the connections between evolutionary theory and German military ideology? And who could say that America was not headed on the same destructive path? With the future of democracy at stake, the evangelical crusade launched its attack after the war on what it perceived as the "anti-democratic, 'might is right,' Bible-denying philosophy of evolution."[69]

The anti-German, anti-Darwinian war rhetoric advanced both by American biologists and by fundamentalists highlighted the need for an evolutionary philosophy in harmony with democratic ideals. Responding to the fundamentalist outcries after the war, American biologists such as William Patten, Edwin Grant Conklin, David Starr Jordan, and William Ritter maintained that the crisis of civilization was not the result of evolutionary doctrine per se but was due to a misunderstanding of evolution's true meaning. In their defense of biology and democratic values, these scientists found recourse in the Spencerian themes of specialization and cooperation as companion principles of progress, although they tailored them to meet the demands of a period marked by progressive reform.[70] American society was entering a phase of increased specialization and social fragmentation, but further progress would result only if better methods of cooperation were found. The Dartmouth biologist, Patten, for example, considered "the functional disease of our civilization . . . to be the decreasing co-operation between the very agencies which have produced it."[71] Organizer of the first compulsory freshman evolution course in the country, Patten reminded his students that just as the parts of an organism cooperate to create something larger than themselves, so did "all the different parts and environments of the college work together for its creation." The college was "a great living organism" of which the students and "many others" were "the component parts." And the student would only be helpful to the college if he or she contributed to its greater welfare, to the common good in which they all shared. It was a civics lesson that applied to American society as a whole.[72] In a similar fashion, the Princeton biologist, Conklin, emphasized the need for "a new revolution which will enforce the duties of man as our former revolution emphasized the rights of man."[73] Evolution, viewed as a creative process of cooperation and mutual service, thus had much to teach with respect to the moral conduct of human society.

In invoking the metaphors of cooperation and organic community, these biologists were tapping into the language of progressive reformers that, at one level, had its roots in the nineteenth-century social gospel of Protestantism. Because the "Progressive movement" never represented any coherent or unified movement as such, American historians have become uneasy with the term's use.[74] But, as Daniel Rodgers has argued, Progressives did draw on a cluster of ideas with distinctly different languages to articulate their visions of social reform. One strain of Progressivist thought drew on the rhetoric of community and social bonds, a rhetoric that a number of historians have traced to a Christian social ethic.[75] The sacrifice of self to community, of cooperation between individuals and institutions, represented in its fulfillment the teachings of Christ symbolized within the democratic order. As a young John Dewey remarked, "the next religious prophet who will have a permanent and real influence on men's lives will be the man who succeeds in pointing out the religious meaning of democracy, the ultimate religious value to be found in the normal flow of life itself."[76]

In this language of community and social bonds, "social Darwinism" emerged as the enemy, or perhaps the straw man of Progressive liberals preaching an American rhetoric of social reform.[77] "Social Darwinism" was supposedly symbolic of an older economic order in which doctrines of laissez-faire and "survival of the fittest" were used as a justification of traditional business ideology of the Gilded Age. In early twentieth-century America, however, the old order based on rugged individualism and competition had passed. Biologists and social reformers were united by their shared use of "social Darwinism" to further their own political ideologies. The attack on the Darwinist origins of German militarism, for example, helped highlight the social values already embedded within certain strands of evolutionary thought. And yet, neither biologists steeped in developmentalism nor Progressives rejected evolution; what they denied was an emphasis on individualism and struggle as part of the evolutionary process. Dewey's philosophy, for instance, relied heavily on evolutionary concepts. The notions of growth, process, and activity, as discussed in Chapter 2, were as central to Dewey as they were to American embryologists such as Patten, Conklin, and Whitman.

Robert Crunden has argued that such new professions as social work, law, journalism, and politics became, in the early twentieth century, the preferred vehicle for preaching social and moral reform. Although Crunden has pointed to the importance of the progressive influence in the social sciences, he claims that within biology, "the progressive influence is visible but basically insignificant."[78] In fact, however, the emphasis on social reform, rational planning, and the professional expert was expressed in a number of guises within the biological sciences. One favored avenue of social reform advocated by American biologists was that of eugenics—a movement that, as Garland Allen argues, was "based on the concept of rational scientific planning in the cause of national efficiency."[79] And apart from the physical and mental improvement of the human individual, biologists also saw their science contributing to the progress of human social evolution. Just as the social science professions perceived themselves as experts crucial to the functioning of a new democratic social order, so the biologist similarly envisioned an exalted position for his own profession. The message of the biologist, Patten told fellow members of the zoology section of the AAAS in 1919, was the message of the social prophet. In his vice-presidential address, Patten stressed, like many others, the importance of cooperation in evolution and biology's part in the war. But he also emphasized the biologist's future responsibility, as one who could liberate the masses from their "ghost-hunter's paleolithic mental attitude toward natural phenomena, and their leaders' . . . similar attitude toward social problems."[80] The religious prophet of John Dewey was, in Patten's eyes, the biologist. For within the sanctity of biology, one found the laws of growth. And through the "knowledge of a living, growing world" came a "broader and a more sympathetic democracy, a deeper love of Nature in all her phases, and a greater assurance of the equality of mutual rights and mutual obligations."[81] This emphasis on the moral teachings of biology and its significance for democracy was an attempt to assert the primacy of biology as a beacon for the conduct and future growth of American society at a time when the cultural authority of biology was itself being challenged. Furthermore, by creating a biological morality congruent with a Christian social ethic, Patten, Conklin, and others hoped to deflate the alleged war between science and religion that surfaced during the 1920s.

Allee was among the audience that listened to Patten's 1919 AAAS address, and at the same meetings, he gave his first paper on the nature of animal aggregations, a subject that would occupy thirty-five years of future research. Patten's cooperationist writings, with their links to Spencerian evolutionism in the vein of Espinas and Kropotkin, appealed to Allee on a number of counts. First, the developmental metaphor implicit in this social evolutionist literature was one familiar to Allee from his graduate days at Chicago. Succession was patterned on the same recapitulationist model of individual growth that informed Spencer's evolutionary views. The development of the ecological climax community was a story of increasing complexity and interdependence, just as the progress of human society was moving toward increasing specialization and cooperation. Second, the use of organicist analogies underscored a physiological approach to studying the individual body and the body politic. Organic functionalism, with its focus on how individual parts were integrated and organized into purposeful wholes, formed the core of the interactionist paradigm in the life sciences at Chicago, a paradigm in which physiology served as the model science. Allee's training in physiology and behavior thus made him receptive to these organicist views. Finally, the emphasis on cooperation and community in these social evolutionist writings, with their explicit ties to a Christian social ethic, harmonized with Allee's Quaker upbringing and his experiences as a liberal pacifist during the First World War. Exposed to this tradition of antiwar biology, and shaken by his own experiences in World War I, Allee embarked on a research program that melded ecological science and politics into a vision of a cooperative world.

4

Cooperationist Beginnings

Today, population ecology represents a distinct branch of ecological science. In the early 1920s, however, the mainstream of ecological research centered on either autecology, with its focus on the physiological response of individual organisms, or on community analysis. Notably absent in animal ecology texts of the 1920s is any discussion of the population as a distinct organizational level with its own properties and phenomena in need of explanation.[1] The extent to which ecologists touched on animal populations came primarily from an interest in animal densities that characterized the work of natural history census surveys, economic biologists, and human demographers. Even Charles Elton's discussion of animal numbers in his influential 1927 text, *Animal Ecology,* attacked the problem from a community perspective. Elton studied fluctuations in animal abundance to unravel the important food relationships that integrated the community into a functional whole. Only in the 1930s, with the development of mathematical population ecology and genetics by theoreticians such as Alfred Lotka, Raymond Pearl, and Sewall Wright and increased interest in the laboratory investigation of the physiology and growth form of experimental animal populations, did varied approaches to the study of population ecology coalesce into a distinct field.[2]

Allee's early career illustrates the existing dichotomy between autecology and community studies that marked animal ecology research. Through the use of chemicals such as potassium cyanide, the same experimental techniques used by Child, he continued to explore the physiological response of individual organisms in the laboratory by investigating the relationship between rheotactic response and metabolic rate.[3] In addition, as his mentor Shelford had done, he extended his analysis beyond the domain of physiological life histories to include research on animal communities. Through his association with the invertebrate course at the MBL, Allee published a detailed investigation of the limiting factors governing the distribution and succession of littoral invertebrates in the Woods Hole region. In addition, he spent time in such distance places as Barro Colorado Island in the Canal Zone pursuing community studies of the tropical rain forest. He drove spikes eighty feet up a sand-box tree and was thus one of the first to climb into the upper regions of the forest canopy in order to analyze the vertical stratification of the rain-forest community.[4]

In the summer of 1917, while the first American soldiers were being drafted into the war, Allee began research on a new line of inquiry that explored the middle ground between autecology and community studies. This work focused on the causes and significance of animal aggregations and represented an attempt by Allee to bring community analysis within the laboratory domain. Because aggregations represented "intermediate units between individual animals and the larger ecological community," they presented methodological opportunities not possible in either individual or community ecology.[5] Through aggregation research, Allee hoped to unravel the significance of interphysiological interactions. Little was known about the nature of biotic interactions that structured animal communities. While most ecologists investigated the impact of the physical environment on the distribution of animal life, Allee was intent on discovering the effects group life had on individual physiology and behavior. Such investigations, he believed, would, yield insights into the problems of community integration, the importance of biotic relations, and the origins of sociality. At the aggregation level, the same forces that operated within the community were present, but they acted in a much simpler and more accessible way.[6]

Although Allee pursued his experimental research within the framework of community ecology, his interests in animal aggregations had a much broader significance. Animal aggregations were not only important for understanding the nature of animal communities, they also formed the cornerstone for a general theory of sociality. Indeed, Allee's 1931 book entitled *Animal Aggregations: A Study in General Sociology* outlined the "border-line field where general sociology meets and overlaps general physiology and ecology."[7] Animal communities and human communities were part of the same developmental processes, governed by the same natural laws. Thus, animal aggregations had direct relevance for human society, and it was the ecologist's "responsibility for interpreting [this] scientific point of approach to human problems."[8] In postulating a grand unifying theory of sociality, Allee reflected the belief, shared by other biologists of his generation, that a biological humanism could uncover the source in nature from which human values sprang and reveal the future direction of human social progress.

The Benefits of Community Life

While Allee advocated a comparative approach to the study of animal and human sociality, sociologists were busy erecting their own disciplinary boundaries, arguing for the importance of human society as a distinct academic subject. As one sociologist suggested, "Wherever there is group life there are problems that he [the biologist] has no technique for studying."[9] From the biologist's perspective, however, human society was but one instance of social life in the animal kingdom. During the interwar years, a number of biologists conducted research on animal societies with the implicit assumption that these investigations bore directly on problems in human sociology. William Morton Wheeler, a professor of economic entomology at Harvard and prolific writer on ants and the social insects, exemplified this position when he remarked that "the biologist will look askance at all the attempts of the ideologists to sever, or even to stretch unduly, the bonds between his science and sociology."[10]

Of all the biological disciplines, ecology contributed the most to this comparative sociology. Plant and animal communities were simply one form of association, and ecology texts such as Josias Braun-Blanquet's *Plant Sociology,* a text that Allee used in his animal aggregation course at Chicago, were laden with sociological metaphors.[11] Allee's own studies were pursued within this general context. He had, as early as 1923, laid the interpretive framework for this research. For Allee, a continuity existed between the loosely integrated associations of the ecologist and true societies. Sociality, he argued, was not to be derived from the family but instead arose from "the consociation of adult individuals for cooperative purposes." Human society was but an extension of the associations that existed throughout the animal kingdom—associations that arose as a consequence of the mutual benefit group life provided in the individual's struggle with its environment. Spencer's theory on the origins of society thus had special appeal because Spencer stressed "the importance of the gang rather than the family as a preliminary step in the evolution of the social habit."[12]

Many aggregations, Allee noticed, arose from either individual tropisms or trial and error responses. The water isopod, *Asellus communis,* for example, formed bunches during the breeding season as a result of a strong reaction to touch. Similarly, in land isopods, aggregations were commonly formed in response to dessication, while in the brittlestar, *Ophioderma brevispina,* groups arose as a consequence of individual reactions to light.[13] In each case, the aggregation formed on the basis of individual behavioral reactions to surrounding environmental conditions. In this respect, Allee's ideas closely mirrored the position of social nominalism embraced by Spencer, a position that dominated American sociology well into the 1910s and one reflected in the pragmatism of Dewey.[14] Sociality, Allee argued, arose solely from individual behaviors and interactions; no social instincts were involved. "The only social trait necessarily present," he wrote, "is that of toleration for the presence of numerous other similar animals within the region."[15] As a graduate student under Shelford, Allee became acquainted with the writings of the French crowd psychologist Gabriel Tarde which reinforced the individualistic strand at the heart of the nominalist's stance.[16] In an article entitled "Inter-Psychology," an article that clarified Shelford's definitions of interpsychology and intermores psychology, Tarde emphasized the distinction between the physiological psychology of the individual and interpsychology, the latter referring to the influences of individuals of the same species upon one another. Tarde rejected the expression "social psychology" because it implied that a social self existed independent of, and distinct from, the individuals composing the group. Reacting against the Comtean tradition of social realism, Tarde suggested that social life was a process that originated from the interactions among individuals. There were no autonomous social facts.

Allee believed the "first step toward the development of social life" was in the "appearance of a physiological tolerance for the presence of other animals in a limited area where they have collected as the result of their individual reactions" to environmental conditions.[17] Without some beneficial value, however, the aggregation had little significance; it was merely a crowd with no cohesive force that bound individual members into a united group. But if sociality had its origins in animal associations, as Allee suggested, then the group must confer some benefit on its individual members if it was to persist for any length of time. By studying the survival value of grouped versus isolated individuals under varying environmental conditions, he hoped to shed light on the nature and significance of group life.

Allee approached the study of aggregations utilizing the research protocol already established for analyzing the physiological life histories of individual organisms. His early aggregation research compared the effects of groped versus isolated organisms on metabolic rate. By holding all physical factors constant except for the number of individuals, he analyzed the effect of group membership on individual physiology, measuring such factors as respiration, water loss, and time till death. In land isopods, for instance, isolated individuals exhibited greater water loss over long periods compared to grouped individuals. Similarly, during periods of starvation, bunches of brittlestars survived longer than their isolated kin.[18] Aggregations thus served as an adaptive physiological response to changing environmental conditions. The situation, Allee suggested, was "similar to that which causes the collection of foreigners into communities in our large cities; that is, a group of similar animals tend to minimize for each other the disturbing effects of unusual surroundings."[19]

Further research on animal aggregations was stimulated by the reports of Georges Bohn and Anna Drzewina in France that grouped individuals of the species *Convoluta* (a marine turbellarian), *Rana fusca* (tadpoles), and various infusiorians survived longer than isolated individuals when subjected to toxic reagants. Drzewina and Bohn attributed this result to the secretion of an autoprotective substance by the individual organisms, a substance that occurred in higher concentrations within the group.[20] A number of errors were apparent, however, in their experimental technique. Because they exposed both grouped and individual organisms to the same volume of toxic solution, the amount of toxic substance *per individual* was much greater in the isolated cases. Consequently, one colleague suggested that the fixation of the toxic substance in the organism's tissue was a more probable explanation. With more individuals, the amount of toxic substance in the environment would decrease. Drzewina and Bohn denied this possibility.[21]

Allee confirmed Drzewina and Bohn's evidence in group survival, but he disagreed with their causal explanation. If the concentration *per individual* of colloidal silver was held constant, the difference in survival value between grouped and isolated individuals was negligible. Furthermore, Allee's experiments indicated that grouped animals actually removed the colloidal silver from solution. No evidence for an autoprotective substance existed. He suggested instead that group protection was "largely and perhaps completely, furnished by the fixation of the toxic substance by the mass of animals."[22] In the case of *Planaria*, the presence of slime functioned as an adsorptive agent. In addition, experimental evidence indicated that protection against colloidal silver was not species-specific as Drzewina and Bohn had suggested. When Allee placed a single *Planaria* in an aquarium with different species such as *Cladocera, Asellus,* or pond snails, the *Planaria* lived longer than it would normally under isolated conditions. In this instance, a multispecies community had definite survival value when survival was measured in time until death.

Exposure of the marine turbellarian, *Procerodes,* to hypotonic sea water furnished another opportunity for examining the French couple's claim of autoprotection. Drzewina and Bohn had noted that individual turbellarians were much more sensitive to hypotonic sea water than grouped individuals. Once again, they postulated the existence of some autoprotective substance secreted by the group, but they did not investigate the nature or origin of this substance. Dissatisfied with this explanation, Allee was determined to find the underlying cause. Experimental work on the adaptation of *Procerodes ulvae* to fresh water by a number of investigators indicated that *Procerodes* had a greater resistance to fresh water when the calcium content of the water was high. Capitalizing on this research, Allee speculated that calcium was the factor responsible for Drzewina and Bohn's observed results. Assays of the *Procerodes*-conditioned water revealed high calcium concentrations compared to normal fresh water. By decreasing cell permeability, calcium reduced the amount of water uptake, extending the life of the organism in a hypotonic environment. Allee agreed that the group secreted an autoprotective substance, but in the case of *Procerodes,* the substance was "nothing more mysterious than calcium."[23]

Placed within the larger context of community studies, these results suggested the existence of integrating factors in animal communities that were more subtle than the traditional food relationships. If, as many ecologists agreed, communities existed as closely integrated units, certain factors were obviously important in uniting the different members into a functional whole. In *Animal Communities in Temperate America,* for instance, Shelford pointed out the importance of food relations in maintaining community equilibrium, and he drew general diagrams of aquatic and terrestrial animal communities showing these connections. The significance of food relationships in community organization was also the central theme of Elton's *Animal Ecology.*[24] But Allee argued that experimental work on animal aggregations "warrants the postulation of another and more subtle integrating factor in animal communities. . . . This integrating factor might well be called the auto-protective value of the community."[25]

Proof that group survival value was not species-specific lent further support to the hypothesis that group protection was of importance to the ecological animal community. The aggregation helped condition the environment, making it more suitable for other forms. "We are proposing an extension of this principle," Allee wrote, "to include the idea that the pioneers in a succession first react upon their environment so as to render it more favorable for supporting themselves and their own associates." He had arrived at experimental evidence for the importance of biotic conditioning, thereby contributing to an understanding of the underlying processes of succession.[26]

Yet a much deeper ethical commitment lay beneath Allee's animal aggregation research. By 1928 Allee felt he had enough experimental evidence on group survival value to support the suggestions of Espinas, Kropotkin, and Patten that cooperation was a fundamental biological principle in nature. "Many of the beneficial effects of relatively unorganized aggregations of animals," he wrote, "are the expression of a vague, unconscious mutual co-operation and . . . the principle of co-operation should rank as one of the major principles comparable with the better recognized Darwinian principle of the struggle for existence."[27] In making an inference from his experiments on group survival value to the principle of cooperation, however, Allee followed the path of Ellen in *Jane's Island.* He had jumped far in one leap, connecting human motives and emotions with a phenomenon that was often purely chemical in nature. One of the same criticisms launched at sociobiologists could apply equally to Allee's work. He united animal and human behavior through a common word, eschewing the different historical and, in the case of humans, cultural context in which each occurred and reduced them both to a single biological cause. Raymond Pearl was one of Allee's few peers to actually criticize him for the nature of this argument. "Does the mere fact that *n* animals survive longer in an unfavorable solution than do $n + m$ or $n - m$ animals, in itself demonstrate that there is any element of true cooperation, automatic or otherwise?" Pearl asked. "Is not further evidence, and of a qualitatively different sort, necessary to prove such a conclusion? . . . It would seem that the sort of observed biological effects . . . here under discussion," Pearl reasoned, "do not, in cold logic, warrant conclusions about cooperation as a fundamental biological principle."[28]

Despite Pearl's criticism, Allee felt he had demonstrated the existence of cooperation at the subsocial level, and, furthermore, had identified an important mechanism in the evolution of social life. The development of society was not confined to the family but had its evolutionary origins in the protocooperative impulses evident in the beneficial survival values of animal aggregations.[29] Indeed, Allee argued, "the potentiality of social life is inherent in living matter, even though its first manifestations are merely those of a slight mutual interdependence, or of an automatic co-operation which finds its first biological expression as a subtle binding link of primitive ecological biocoenoses."[30] For Allee, sociality began with the protocooperative tendencies already found in the most primitive single-celled organisms.

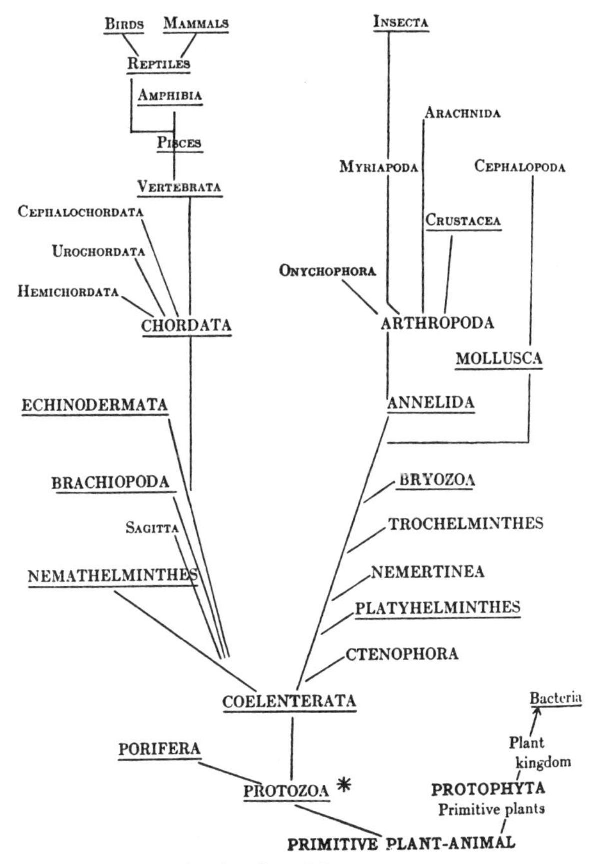

Figure 4.1. Diphyletic tree showing the evolution of the major phyla within the animal kingdom. Phyla and classes underscored indicate the presence of beneficial survival values found in animal aggregations obtained through experiments in Allee's laboratory. From W. C. Allee, "Cooperation among Animals," *American Journal of Sociology* 37 (1931): 390.

In support of his idea that sociality began very early in the evolution of life, Allee often presented a diphyletic tree of the animal kingdom, splitting at the Coelenterates and branching off with humans at one apex and insects at the other (fig. 4.1). Because both insects and mammals have developed closely knit social groups, he inferred that an ancestral origin of sociality extended as far back as the Protozoa. Much of his research program can be seen as an attempt to provide experimental evidence that each of the phyla in this evolutionary tree exhibited protocooperative tendencies. As late as 1949, he attempted to demonstrate beneficial effects of groups in rotifers: one of the few phyla where such effects remained unknown.[31]

Like the principle of succession, Allee's theory of sociality was couched within a developmental view that stressed continuity and progress and harked back to the earlier ideas of Spencer and Espinas. Espinas drew no distinction between the behavior of cells forming an individual organism and the interactions of individuals that make up a society. Paul Deegener's writings appealed to Allee for similar reasons. Deegener, a professor of zoology at the University of Berlin, published, in 1918, an exhaustive classification of animal groups found in nature. Accidental associations, Deegener argued, were either intraspecific or interspecific groupings that had no particular value. In essential societies, individual members benefited from the association. By linking accidental associations and essential societies together in a single classification scheme, however, Deegener drew "no hard and fast line . . . between social and infrasocial" life. In each instance, the transition between social levels was one of degree not kind.[32]

In dismissing sex and the family as the fundamental origin of animal and human society, Allee found himself on the periphery of biological/social discourse which was dominated by anthropological theory in the 1920s. Wheeler, for example, criticized attempts such as Allee's to link the primitive associations of ecological communities with eusocial life. "Some authors," Wheeler wrote, "have endeavored to derive the societies from the associations, but it is difficult to find any cogent proof of their contentions. The societies really represent very different emergent levels from the associations and have arisen in a different way." True societies, Wheeler argued, unlike associations, had arisen from heterosexual relations. The family was a crucial step in the development of social life. "All the societies of insects," Wheeler insisted, "are merely single families in origin." The societies of insects, the flocks of birds, the herds of mammals, and the tribes of humans were all extensions of family life in which the bonds between parents and offspring ensured some degree of social cohesion.[33] Sex and reproduction also structured the major narrative frames in the emerging field of primatology. In a 1928 article, the mammologist Gerrit S. Miller suggested that primate societies were characterized by a great degree of sexual promiscuity due to the absence of an estrus cycle, thus enabling females "to accept the male at all times." This contention formed the basis of Solly Zuckerman's 1932 book *The Social Life of Monkeys and Apes* and pervaded writings on animal sociality throughout the interwar years.[34] From this sexual chaos, the human nuclear family arose, giving birth to culture, superimposing order on the natural world. Sex was an important integrating force in society. As Donna Haraway notes, the nuclear family, based on heterosexual relations, served as a social model in which male dominance created an organic hierarchy that minimized conflict and promoted cooperation among members of the family unit. One could similarly apply this organic model to the workplace, where a newly founded managerial class sought to reduce labor conflict, promoting efficiency and maximization of profits, which, in turn, enabled the company to better compete with other firms.[35]

Allee did not deny that sex was an important element in the structuring of many mammalian societies. His 1931 book *Animal Aggregations,* for instance, contains a lengthy discussion of Miller's article and outlines many instances in which he considered sex to play an important role in social life. What Allee denied was that *all* social life emerged from reproductive relations. He insisted that a model of society other than the family could be found in the animal kingdoms; this model was the animal aggregation or association, where individuals gathered in response to physical conditions in the environment. Not integrated by sex, the aggregation was a community of organisms bound together through a common struggle with their physical surroundings.

The Politics of Sex

Allee's rejection of the family as *the* nascent society was not unique. His anthropologist friend Ralph Linton expressed similar views in *The Study of Man,* published in 1936. Trained under Franz Boas, Linton achieved professional recognition for his dissertation on *The Material Culture of the Marquesas Islands.* In 1922, he accepted a position as assistant curator at the Field Museum of Chicago but moved on to the Department of Sociology at the University of Wisconsin in 1928.[36] It is unclear how Linton and Allee first met.[37] Allee had established connections with the Field Museum in the 1920s, primarily through K. P. Schmidt. Shortly after Schmidt moved to the Field Museum in 1922, he approached Allee with the idea of translating Richard Hesse's *Tiergeographie auf Oekologisher Grundlage.* Schmidt thought that such a translation could count toward an advanced degree under Allee and would also lay the groundwork for a modern text on ecological and historical zoogeography which Schmidt hoped to write.[38] Allee agreed to participate in this collaborative venture, and Schmidt began the task of translating in 1925, but the actual volume, *Ecological Animal Geography,* did not appear until 1937 due to difficulties in securing a publisher. The volume was significant because it made available in English one of the first books, apart from Shelford's *Animal Communities in Temperate America*, to offer an account of the worldwide distribution of animal life on a physiological as opposed to historical basis.[39] This joint venture opened the way for other collaborations between Chicago's zoology department and the Field Museum. In 1931, for example, the museum gave the Department of Zoology access to collections and exhibits for its courses.[40] It is thus possible that Allee met Linton through these associations.

The Hyde Park group of the Chicago Monthly Meeting of Friends was another likely spot where Allee and Linton might have crossed paths. In the 1920s, a strong core of Quakers affiliated with the University of Chicago met on a regular basis, often at the home of Paul Douglas, an economist at Chicago who later became city alderman and state senator. In 1931, this group officially became the 57th Street Meeting of Friends. Active in social causes such as labor relations and interracial housing, the group sponsored regular forums to discuss social and political issues of the day, drawing speakers from the university faculty, Chicago settlement houses, and other social reform movements.[41] Linton was himself a birthright member of the Society of Friends and, like Allee, attended Quaker educational institutions through high school and college, enrolling at Swarthmore in 1911. He did not, however, share Allee's pacifist convictions and enlisted in the summer of 1917. Linton found himself in the Rainbow Division, one of the first American units to see active combat on the French front. Yet the shared Quaker background of Allee and Linton, with its disdain for authoritarianism and respect for individual values, is suggestive of the common context in which they developed their social theories.

In *The Study of Man,* Linton suggested that anthropologists had paid too much attention to the role of the family in the development of social life, a preoccupation that he believed stemmed from European culture's "extreme interest in everything connected with mating and reproduction."[42] While anthropologists like Bronislaw Malinowski had pointed to the "family as the cradle of nascent culture," Linton argued that the "local group . . . has been as important as the family in the development of social institutions."[43] Both the family and the group were "raw materials" upon which society was built. Army units and lumber gangs were but a few examples where unrelated individuals gathered together on their own accord. The necessary condition for society to emerge was an aggregate of individuals that persisted over time; kinship was not essential. Once the aggregate continued over an extended period, "integrating forces" would "have an opportunity to act." In the case of the family, the biological demands of sex and care of offspring ensured some degree of continuity. Linton envisioned that the first human beings "lived in male-dominated and frequently polygynous families."[44] In this respect, he did not break from the patriarchal assumptions underlying organic functionalism of the period. But Linton also believed that these families were at the same time organized into local groups or hordes in which "no one" individual is "leader." Persistence of the local group was "ensured by habit," by the conditioning of individuals to each other's presence. For Linton, the local group had an extremely important social function. It presented the "optimum conditions for the transmission of culture" and created a stable social environment with the minimum of formal government or authoritarianism. He lamented the demise of this organizational structure with the rise of industrial civilization. Masses of individuals groped about in the modern city searching for their lost bands. The future, Linton feared, resided in the totalitarian state, where all sense of individual expression and freedom was lost.[45]

In Linton's view, the transformation of aggregates into societies occurred as the result of two processes. The first was "adaptation and organization of the behavior of the component individuals." Linton placed a high value on the individual. He adopted a social nominalist position, dispensing with any notions of a "group mind or a group soul." Society, he insisted, "is a group of biologically distinct and self-contained individuals whose psychological and behavioral adaptations have made them necessary to each other without obliterating their individuality."[46] Once individuals gathered together, a period of adjustment followed in which group members adapted to one another, each assuming a certain role, specializing in a certain task, thereby reducing potential conflict so that the group could work effectively together. Yet society did not fully emerge until "a community of ideas and values" was established, an "esprit de corps," a sense of group consciousness, where the individual was "willing to sacrifice his own interests to those of the whole."[47] What distinguished society from an aggregate, Linton argued, was its "organization of mutually adapted personalities." Integration took place within society at the psychological level. Eventually, a set of idealized patterns formed that organized and constrained individual interactions, patterns that Linton believed had a "superindividual character." Even though society arose through individual interactions, once formed, the group possessed properties that could not be discerned solely from a study of individual behavior.[48]

Linton's portrayal of social evolution broadens the context and meaning of Allee's aggregation research. Indeed, one can trace the direction of Allee's experimental program according to the development of society in Linton's terms. In the 1920s, Allee established the physiological conditions by which animal aggregations first formed. He thus provided biological evidence demonstrating the existence of animal associations formed on the basis of individual reactions to the environment. Yet, for sociality to develop, the aggregation had to persist in time; tolerance of other individuals in the group was not enough. Allee's experiments demonstrating group survival value offered an explanation for the continued presence of the group: the aggregation offered benefits unavailable to the individual pursuing a solitary life. In the 1930s, Allee moved beyond the study of integrating mechanisms at the physiological level. He began to explore psychological interactions such as leadership, territoriality, and dominance-subordinance relations that structured social organizations in the higher vertebrates. Here, however, the problem of sex entered in, presenting a serious challenge to his aggregation work. Indeed, the study of sex threatened to undermine the harmony between his biological and political views, a tension addressed in chapter 8. Why was sex so important? Why did Allee offer an alternative model of society, a model in which sex played an insignificant role?

The birth of society through the family posited a sexual division of labor, a patriarchal hierarchy in which the male, bearer of culture, ensured stability and order by keeping nature in her place. The female, constrained by biology and thus bound to nature, was destined for a life of reproduction and child rearing. This hierarchical vision, however, conflicted with Allee's political stance. He sought a biological model of society that was explicitly nonhierarchical, one that legitimized the politics of liberal pacifism and his own Quaker beliefs. Through his aggregation research, Allee had found experimental evidence against the doctrine of war, and his emphasis on the principle of cooperation was interdependent with his own experiences as a pacifist during the First World War (fig. 4.2). Although one strand of antiwar biology focused on war's dysgenic effects, eugenic arguments offered no guideline or ethic for political action other than neutrality. Jordan and Kellogg, the most vociferous opponents of war on eugenic grounds, were part of an earlier conservative faction of the American peace movement. But Allee's affiliations with the liberal pacifist movement through the American Friends Service Committee emphasized peace as a political exercise. Cooperation and nonviolence were tools of this political process; power was to be shared equally among individuals or groups of the commonwealth, a commonwealth that would someday be international in scope. Recall that for the liberal pacifist, society was ordered for the benefit of its individual members, yet each member had his or

Figure 4.2. Three species of ants, normally antagonistic, living harmoniously together after the larvae were reared together. W. C. Allee, "Cooperation among Animals," *Chicago Alumni Magazine* (1928), frontispiece.

her own social responsibility to fulfill. Consider Allee's interpretation of animal aggregations. The aggregation provided benefits to each member of the group, but the value of the group was itself the result of individual responses. By contributing to the larger social welfare, each individual profited from the exchange. Furthermore, the process of group formation was the result of consensus rather than domination or control. No leadership or authority existed within these animal aggregations. The group formed as the result of individuals reacting to their physical environment. Once the group formed, each group member displayed a greater physiological tolerance of other members. The power of the group was not based on a hierarchical organization but was distributed equally among individuals. Still working within an organic, physiological metaphor, Allee advocated a politics of peace, not social control.

Much of the experimental methodology Allee utilized to explore the biological meaning of this organicist trope, he learned from his colleague and teacher Charles Manning Child. Yet Child, although adopting the same tools of investigation, turned them to craft a different social tableau. His work in physiological morphology greatly influenced the interactionist, behavioral orientation of animal ecology research developed at Chicago. When Allee studied animal aggregations, he attacked the problem from the standpoint of group physiology—his interests centered on the integrating mechanisms and processes that contributed to the survival and maintenance of the group. Similarly, Child's work on regeneration emphasized the priority of function over form in the organization and maintenance of the organism as a whole. The establishment of the journal, *Physiological Zoology,* at the University of Chicago in 1928 accentuates their shared interests. Although the journal was initiated by Child, Allee served on the first editorial board and took over as managing editor upon Child's retirement in 1934.

From Child's perspective, both the organism and society represented a dynamic order formed by the integration of living parts. One could "expect, therefore, to discover a fundamental similarity . . . in the more general laws and processes of integration from the one extreme of . . . the simplest organism to the other of the great modern state or nation."[49] Each organization developed integrating mechanisms as the result of reactions with its surrounding environment. Hence, a physiological continuity existed from the simplest organisms to the most complex societies. Beyond this organicist approach, however, the agreement between Child and Allee on social organization dissolved. Child incorporated the concepts of physiological dominance and subordination into discussions of social progress. Social integration was dependent on some degree of leadership and, hence, dominance for its development. "In any persistent orderly integration such as the state," Child argued, "a definite and persistent relation of dominance and subordination exists and on it the orderly character of the state depends."[50] Primitive autocracy characterized the lower levels of organization in Child's social stage theory. The dominant region controlled the subordinate parts through coercion rather than cooperation or consent. Further progress toward a physiological democracy depended on the development of leadership into a highly specialized and efficient type, such as the nervous system in the higher organisms, and on the development of better means of transmission or communication.

But the nervous system of higher organisms was not, Child envisioned, an autocracy. The subordinate regions signaled back to the nerve centers of the brain, participating to some extent in directing the organism's future activity. Competition was an important part of this interactive activity. In the model democratic organism, Child argued, "there is governmental control, regulation and coordination of the various parts, but there is also conflict and competition."[51] Only through inequality and conflict could patterns of domination and subordination be established, which, in turn, integrated the individual members into a functional whole. Dominance, hierarchy, and the division of labor, the pillars of organic functionalism, reverberated throughout Child's views.

From Allee's standpoint, dominance did not bring order but war. By the early 1930s, he had developed a research program within the domain of animal ecology that denounced war on biological grounds and placed the politics of peace on a naturalistic foundation. Yet he still pursued his aggregation research within a community framework. His theory of social evolution mirrored that of ecological succession. Both were developmental metaphors of progress that discussed evolution without any reference to hereditary mechanisms. With the rise of Mendelian genetics and the popularity of eugenics, however, the study of heredity occupied the limelight of biological investigation. And the site to which biologists could direct their weapons to cure social ills resided in the race, in the population. Although the appeal of eugenics within the United States had faded by the early 1930s, still the population occupied a central place in biology. As a genetically distinct group of individuals, the population was the evolutionary unit upon which both the biologist and natural selection could act. Over the course of a decade beginning in the mid-1920s, theoretical developments in the fields of genetics, biometry, and demography, in addition to field and experimental work on animal numbers, helped crystallize the population as a distinct entity worthy of analysis. During the 1920s, Allee was somewhat isolated at Chicago. In Schmidt, he found another naturalist; yet someone who was trained in evolutionary rather than physiological science. The addition of Sewall Wright and Alfred Emerson to the zoology department at Chicago in the late 1920s also brought new perspectives different from the organicist, physiological ideas regenerated in the department over the previous twenty years. Emerson and Wright were educated, as Schmidt was, in an evolutionary viewpoint directly informed by developments in the study of genetics. Their influence, along with other institutional changes, prompted Allee to retool his ecological ideas in the 1930s. He partially abandoned the successional model of social evolution. Having to contend with a hereditary mechanism for the evolution of cooperation, he incorporated a group selectionist model, in which the population was the fundamental unit of evolutionary change. With the addition of his student Thomas Park to the staff in 1937, a core group of animal ecologists emerged at Chicago, united along a common conceptual front that emphasized the study of the population as a distinct physiological and evolutionary unit. And reflecting the sentiments of biologists elsewhere, they, too, saw the population as the site for social change.

5

Population Problems

While biologists such as Kellogg implicated German evolutionary theory as a cause of the First World War, others interpreted the war situation as a neo-Malthusian problem of overpopulation. Indeed, the problem of population became a topic of central concern after the war, especially among social scientists and biologists trained in statistics and demography. Raymond Pearl was one such biologist seasoned in quantitative methods who became interested in questions of population growth in the early 1920s. Pearl received his Ph.D. at the University of Michigan in 1902 and spent a year abroad with the British biometrician Karl Pearson in 1905–6. From 1907 to 1917, Pearl worked at the Agricultural Experimental Station at the University of Maine. With the United States' entry in the war, he was called into government service, acting as chief of the Statistical Division of the United States Food Administration. In 1918, he left Washington for a position as professor of biometry and vital statistics at Johns Hopkins University where he remained until his death in 1940.[1]

In *The Biology of Population Growth,* published in 1925, Pearl discussed why the war acted as a catalyst for a burgeoning literature on population problems. "In the first place," Pearl remarked, "population pressure is always a major cause of war. In the second place, war always disorganizes . . . the pre-existing economic structure." The war created temporary markets that increased employment among the civilian population, but after the war, production demands decreased and the returning troops caused a labor glut that led to unemployment. Hence, a "vague and inarticulate but wide-spread feeling" surfaced "that there are too many people in the world."[2] The war was, however, not the only cause of population concerns. As Congress debated the passage of the Immigration Restriction Act of 1924, attention was drawn to shifting demographic trends in the U.S. immigrant population, a point behind which eugenicists rallied in their attempts to ensure that the Nordic stock of the American population was preserved.[3] The birth control movement, also aligned in its early years with eugenic reform, fed into this discussion of population growth rates among races, classes, and nations.[4] Spurred on by this hubbub of talk about human numbers, the American Statistical Association devoted its 1924 meeting to a forum for papers on population problems.[5] Under the leadership of Pearl, a body of mathematical theory emerged in the 1920s, to analyze and describe population phenomena; one of the most notable developments that had repercussions for experimental animal ecology was Pearl's formulation of the logistic growth curve.[6]

Allee was not oblivious to these population parleys. In 1921, he reflected on the causes of the Great War. Population pressure in Western Europe, he believed, had created a struggle for resources. To avoid war, one needed either to control population size or increase production of foodstuffs and energy reserves. Allee thought the latter could be achieved if one turned tropical rain forests into high-yielding agricultural areas, thereby reducing potential conflicts over shortages of food and other resources.[7] Despite his awareness of the literature on population growth, especially as it related to war, his animal aggregation research in the 1920s was little informed by population theory. He regarded the aggregation as a tool for experimentally analyzing problems in community ecology, especially issues relating to succession and community integration. That the aggregation presented unique questions separate from the community domain was not a primary concern. When he returned to the University of Chicago in 1921, he began rebuilding an animal ecology program that had stood vacant since the departure of Shelford in 1914. This program included a course sequence in field zoology, animal ecology, animal geography, animal behavior, and a graduate seminar in animal behavior and ecology (fig. 5.1). Indicative of his new research direction and the value Allee placed on the study of biotic interactions, he added a graduate seminar on animal aggregations in 1924. By the late 1920s, he started to attract a handful of graduate students, and while a few pursued ecological questions in the field, the majority of dissertations dealt with some aspect of mass physiology in the laboratory (fig. 5.2). In order to support the accumulating numbers of research assistants and expense of laboratory apparatus, Allee soon found himself confronted with the task of obtaining outside funds. His first attempt at promotion took him across campus, beyond the disciplinary boundaries of biology, to the Local Community Research Committee in the social sciences. Here, he first encountered the need to reclothe his aggregation research in population garb.

Figure 5.1. The 1923 ecology class at the Indiana dunes. Allee is standing second from the left. Libbie Hyman, a long-time research associate in the department is seated in the left-hand corner. Courtesy of the University of Chicago Archives.

Exploring the Borderlands

The Local Community Research Committee was an interdisciplinary research group at Chicago drawn from faculty members in political science, sociology, anthropology, history, economics, and political science. The committee's aim was to develop research methods and techniques in the social sciences by focusing on problems of the local community. Studying a common area such as Chicago would, it was hoped, lead to cooperation among the social sciences and cross-cultivation of research methodologies and techniques. Originally granted $21,000 for 1923–24 from the Laura Spelman Rockefeller Memorial, the budget of the LCRC ranged from $80,000 to $100,000 per year between 1924 and 1927. Funds were used to pay for release time from teaching duties, graduate student stipends, and statistical and clerical assistance. Projects in the first three years included sociological research under the direction of Robert Park and Ernest Burgess on urban area spot maps, family disorganization, juvenile gangs, and the distribution of recreational institutions. In political science, Charles Merriam and Harold Gosnell conducted a study of nonvoting behavior among Chicago residents.[8]

In October 1927, as the three-year grant was coming to a close, members met to decide on the future long-term program of the committee. Allee was aware of the generous funds available for social science research at the university, and he was also cognizant of the extent to which Robert Park, one of the influential leaders in the Chicago school of sociology, was borrowing metaphors from plant ecology to enrich his work on the urban ecology of Chicago and surrounding communities. Yet despite Park's ecological interests and his close proximity to the zoology faculty, he rarely cited Allee's work. The metaphors of competition and dominance, which were central to Park's sociological theories, rarely appeared in Allee's early writings.[9] Child's interests, on the other hand, proved valuable to Park because Child stressed not only the importance of function in determining form but dominance as a means of social control.

"In the study of the social group," Park emphasized, "the point of departure is properly, not structure, but activity."[10] Institutions arose from concerted action, from a shared need or activity of people gathered in a common area. Over time, this function became embodied in a particular structure, custom, or tradition, thus taking on an institutional form. In this process of institutionalization, which resulted indirectly from a struggle between competing forces, patterns of dominance and subordination were established. After hearing a talk given by Child to the Society for Social Research in 1926, Park and Burgess incorporated Child's gradient concepts into their own research on city life.[11] Dominance patterns, Park maintained, established by the city's terrain, economic activity patterns, and neighborhood groupings, determined "the general ecological pattern of the city and the functional relation of each of the different areas of the city to all others." Through competition, patterns of dominance and subordination were established, which brought social order to the community. Consistent with the general tenor of organic functionalism, Park considered that the primary significance of dominance was "to stabilize, to maintain order, and permit the growth of structure in which that order and the corresponding function are embodied."[12]

Park's interest in Child's biological ideas, however, seemed to have little effect on the LCRC's consideration of Allee's work. Throughout the summer and fall of 1927, Allee discussed with Merriam the nature of his aggregation research and his hopes for obtaining financial support. When asked by Merriam for suggestions regarding the com-

Figure 5.2. Allee in the Whitman Laboratory, completed in 1926. Courtesy of the University of Chicago Archives.

mittee's overall program, Allee responded that such a project needed "the foundation, or at least the checking-up of biological research on non-human material." He believed his work on animal aggregations showed a "distinct possibility of helpfulness to such a program," and he looked forward to "the chance of co-operating with" Merriam and the committee.[13] Merriam agreed with Allee's assessment of the need for a biologically trained person, but Allee was not the man he had in mind. The committee had already discussed the possibility of obtaining an appointment in "Biology and Social Relations" as one of several in the "cultivation of borderlands" with other fields. Merriam thought the "development of social research in the University" had crystallized the need for a "consideration of basic problems of population which will require the services of someone trained in biology or statistics or both, and at the same time familiar with and interested in some of the basic problems of the social sciences." "The basic facts developed by biological analysis, the relations between the geneticists and the environmentalists, the intimate study of biological differentials with reference to groups, classes, and races, the development of the eugenics movement—all these," Merriam observed, "have a direct relationship to the fundamental problems in social research." "I do not see how we can hope to build up social science," he concluded, "if we neglect these factors now."[14]

Although never hired, Alexander Carr-Saunders was the committee's candidate of choice. Trained in zoology at Oxford, Carr-Saunders studied with Karl Pearson in London in 1910. While in London, he became subwarden of Toynbee Hall and acted as secretary of the research committee of the Eugenics Education Society. After the war, he accepted a position as demonstrator in zoology at Oxford and completed *The Population Problem,* published in 1922. In 1923, he occupied the Charles Booth chair of social science in Liverpool and went on to become director of the London School of Economics in 1937.[15] The notion of the optimum was a central theme in *The Population Problem,* an idea Carr-Saunders gleaned from the British economist Edwin Cannan and that had been introduced into twentieth-century economic theory by the German economist Julius Wolf. Malthus presented overpopulation as a continual threat to society that could only be checked by vice and misery. Carr-Saunders argued, in contrast, that an intermediate equilibrium point existed between underpopulation and overpopulation determined in relation to natural resources around which population numbers fluctuate. In a population of either a minimum or maximum size, the standard of living is low. When a population grows beyond a certain point, for example, given a limited amount of resources and stagnant technologies of production, the per capita product decreases. Similarly, if the population is too sparse, the shortage of labor leads to scarce goods. Population size will thus tend toward an optimum or mean that maximizes the per capita income of consumer goods relative to the available resources. How did this regulation of numbers first arise? Carr-Saunders reasoned that such a situation originated among territorial primitive tribes living in cooperative groups that competed with neighboring bands. Humans differ from other animals, he argued, in their ability to consciously practice birth control. Thus, if one group adjusted its numbers to ensure that the maximum average return per individual was achieved in relation to the land's resources, while another tribe did not, the cooperating tribe would outlast and outreproduce its competitor, or so Carr-Saunders believed.[16] His model anticipated later biological arguments for group selection, which reached their high point in the 1962 book *Animal Dispersion in Relation to Social Behavior* written by V. C. Wynne-Edwards.[17]

The Population Problem points to a number of themes that surfaced in Allee's writings and in the general population ecology literature during the 1930s. The first is the association of best with optimum, implying that an ideal number of individuals exists which the environment can support and that there are mechanisms within the population that regulate its size about this number. The second is the incorporation of group selectionist models to account for the maintenance of optimum numbers. Carr-Saunders's model proposes selection operating between tribal groups, not between individuals of the same group. Because one group is able to regulate its numbers while the other is not, certain properties are assumed to be operating at the group level independent of the individuals composing the group. Third, although not all biologists attached an explicit social agenda to their study of animal numbers, the shadow of human society loomed large backstage. Population cycles of lemmings are far distant from business cycles in the economy, but in a period when large-scale fluctuations in the stock market signaled the economic downturn of the Great Depression, discussions of regulation and control reflected a widespread concern with stabilizing the economy in the hopes of maximizing or at least securing a higher standard of living. In addition, population control could be used as a guise for eugenic policies without explicit allegiance to the eugenic movement, a movement that had come under increasing suspicion and attack by the late 1920s. Indeed, Garland Allen has detailed the easy ideological shift from Raymond Pearl's support of eugenics to population control. The "social value of each of these movements," Allen writes, "was the same: a desire to improve society through the use of known biological principles" determined and set forth by a body of professional experts.[18] As Carr-Saunders noted, quantity and quality were both at the heart of the population problem. The question was not just how many, but how many of whom. This is why individuals like Pearl hoped to arrive at quantitative estimates of fertility and mortality rates among various cross-sectors of the population. If the lower classes were outbreeding the wealthy elite, the birth control campaign needed (supposedly) to reach the urban poor in order to ensure that the genetic strength of the population be maintained.

Amid this social context of population discussions, Allee began to redirect his thoughts regarding his own animal aggregation research. His interactions with the Local Community Research Committee awakened him to the potential population research had for drawing outside financial support. Pearl, for example, received $175,000 between 1925 and 1930 from the Rockefeller Foundation for the creation of the Institute of Biological Research at Johns Hopkins where he pursued his studies on population growth.[19] In the 1920s, reproduction was fast becoming a topic in the forefront of biological research. Pearl approached the study of reproduction primarily from a demographic angle. The establishment in 1922 of the National Research Council Committee for Research in Problems of Sex, however, provided substantial monetary support for research on the physiology of reproduction, on the internal mechanisms controlling sex. Both approaches aimed at control of reproductive processes; the main differences stemmed from the perspectives of the various disciplines involved. Funded by the Bureau of Social Hygiene from 1922 to 1933, the NRC-CRPS received a total of $720,000 over this eleven-year period. $146,000 of this money went to Allee's colleague, Frank R. Lillie; he was the largest single recipient of NRC-CRPS funds during the first eleven years.[20] At Chicago, the population was a center of biological interest, as it was elsewhere. And although the avenue of investigation was physiology rather than mathematics, still, the underlying social motivation of Lillie, like that of Pearl, was tied to population control. Lillie's attempt to establish an Institute of Genetic Biology in the early 1930s, in addition to the hiring of Alfred Emerson, Thomas Park, and Sewall Wright, created an institutional framework supportive of population research, helping orient Allee's aggregation studies in new theoretical directions.

The Institute of Genetic Biology

An adept administrator, Frank Lillie had done much to build the biological sciences institutionally at Chicago since the time he took over chairmanship of the zoology department from his mentor Whitman in 1910. Lillie traversed a difficult road in his first years as department chair. He first undertook the task of revamping the department's undergraduate curriculum, to which little attention had been given under Whitman's reign. To assist in this undergraduate reform, he hired, in 1911, Horatio Hacket Newman, a Chicago Ph.D. who was also Lillie's cousin-in-law. In addition, Lillie united the embryology section with the rest of zoology, for the two had previously been under separate budgets. Although this integrated departmental members, it also enabled Lillie to secure the bulk of departmental funds for embryology. But the transition of power from Whitman to Lillie was not altogether smooth. In the early 1910s, Lillie faced opposition and growing discontent among his staff. As an instructor, Shelford occupied a precarious appointment; he felt himself an outsider in the department and resented taking on extra course loads to build up the undergraduate curriculum when his future was insecure. Most of the dissent, however, stemmed from Lillie's support of William Lawrence Tower, a figure whom the rest of the faculty distrusted with respect to both departmental affairs and zoological research. Newman considered Tower to be the "real underlying trouble in the department." He wrote to his cousin-in-law in 1912 that the only way of strengthening zoology at Chicago was to enlist "the cooperation of Child, Shelford, Strong and me by severing [Lillie's] alliance with Tower." Tower resigned from the university in 1917 due to a divorce scandal and dissatisfaction among the student body. With Tower's resignation and the appointments of Carl Moore in 1916, Albert Bellamy and Benjamin Willier in 1919, and Allee in 1921, Lillie assembled a staff around him generally supportive of his ideals. All of the faculty were, after all, Chicago progeny: Moore and Willier were both trained under Lillie, Bellamy had been a student of Child's, and Allee was himself a Chicago product.[21]

Ensuring cooperation among rank-and-file members of the department was one way of achieving institutional success, but Chicago's stature as a first-rate zoology department was also dependent upon research. One discovery that created a high profile for zoological science at Chicago and enabled it to draw in considerable outside foundation support was Lillie's investigation of the factors responsible for the development of the freemartin, a sterile twin of a normal male calf. Through an analysis of some 55 pairs of fetal twins obtained from the Union stockyards in Chicago, Lillie concluded that the freemartin was a genetic female that possessed certain male sexual characteristics that were caused by the action of a blood-borne chemical substance, presumably a male sex hormone. Chicago became a leading center for investigating the influences of sex hormones on development and isolating testosterone in the laboratory. To undertake the necessary detailed biochemical work, Lillie established an association with Fred Koch in the Department of Physiological Chemistry. With the creation of the NRC-CRPS, outside funds became available for this research. By 1928, Lillie's operation had grown into a full-scale research enterprise; he and Koch received approximately $24,000 per year in NRC-CRPS monies, and as many as a dozen researchers at a time worked in Lillie's lab. In order to meet the increased demands for laboratory space, Lillie and his wife Frances privately funded the construction of the Whitman Laboratory, completed in 1926, where many of the experimental researches of the department became housed.[22]

Lillie had a grand organizational plan for the biological sciences at Chicago, one in which embryology and sex research occupied center stage, but one that drew in other biological sciences and related disciplines such as genetics, ecology, psychology, and physiological chemistry. In essence, he envisioned an interdisciplinary institute for biology similar to the Local Community Research Committee established for social science research at Chicago. Such an organization implied, however, a common problem or focus that could be pursued on a number of fronts. For the LCRC, the local community, the city of Chicago, was the organizational glue holding together the diverse disciplinary strands. What did Lillie see as an analogous problem in the life sciences? The project was a modified version of eugenics that focused on the biological basis of social control. Lillie was a member of the Eugenics Education Society of Chicago, and also served on the general committee of the Second International Eugenics Congress and the advisory council of the Eugenics Committee of the United States.[23] Although born of a modest family, Lillie's marriage to Frances Crane transported him across class lines into the circles of the wealthy elite. He had much to gain in his espousal of the notion that the lower echelons of society not breed like rabbits, for they were the very class that threatened to undermine his own social lot. Writing in the student newspaper the *Daily Maroon,* Lillie justified his plans by reminding readers that if "our civilization is not to go the way of historical civilizations, a halt must be called to the social conditions that place biological success, the leaving of descendants, in conflict with economic success, which invites the best intellects and extinguishes their families."[24]

Lillie's plans for an Institute of Genetic Biology indicate the extent to which eugenics was, in the late 1920s, being transformed into discussions of population problems. "The future of human society," Lillie wrote, "depends on the preservation of individual health and its extension into the field of public health; but it depends no less on social health, that is the biological composition of the population. We are at a turning point in the history of human society—the age of dispersion and differentiation of races is past. The era of universal contact and amalgamation has come. Moreover, the populations press on their borders everywhere, and also, unfortunately, the best stock biologically is not everywhere the most rapidly breeding stock. The political and social problems involved are fundamentally problems of genetic biology." Lillie envisioned a eugenics program not in a strict hereditary sense where the locus of control was the individual germ plasm, but rather a eugenic program on a sweeping front that included biological sciences under Lillie's rubric of racial biology such as genetics, ecology, embryology, and the biology of sex, areas that, in Lillie's words, "furnish the basic scientific foundation for social control." Beneath this program lay the assumption that the social ills of human society were caused by biological problems and could thus be solved by biological means. But Lillie had little appreciation for the rather simplistic approach taken by Mendelian genetics in studying the problems of heredity and development. The approach of his institute thus differed from the eugenics movement in its emphasis on mastering environmental factors and biological processes other than heredity that affected development and behavior.[25]

In June 1924, Lillie wrote to Wycliffe Rose, then president of the Rockefeller-funded General Education Board (GEB), about his interests in racial biology. One of the primary aims of the GEB during the 1920s was to improve scientific and medical education in America by providing large endowments to top-ranking institutions. With GEB money in mind, Lillie suggested to Rose the importance of developing research in genetic biology which included research in the "physiology of reproduction, in the biology of sex, in the physiology of development and in experimental evolution," themes that were already a focal point of biological research at Chicago. Although no specific proposal was mentioned, at the end of his letter Lillie made a plea for the establishment of "institutes of racial biology where the problems of genetic biology and physiology will be studied with the same intensity as problems of medicine in our medical research institutes." Lillie envisioned an institute at Chicago that would become the leading center for research in racial biology just as the Rockefeller Institute served as a model for biomedical research.[26]

In essence, Lillie wanted to establish an institute that would direct scientific inquiry toward social problems of the human race rather than problems of the human individual, as in the case of medical research. The proposal for such an institute did not appear, however, until the summer of 1930. On 23 June, Lillie wrote to the University of Chicago president, Robert M. Hutchins, in regard to the future of the zoology department. Complaining of inadequate facilities, Lillie suggested that "a positive program can be presented in the way of developing the main trend that our work has taken for many years." This trend, Lillie argued, centered on problems of evolution and racial biology. In his proposal, Lillie distinguished between biological research on the individual and on the race. The former, Lillie argued, was closely allied to medical research and its associated fields such as anatomy, physiology, biochemistry, biophysics, hygiene and bacteriology, and pathology. Fields related to racial biology, which included genetics, cytology, embryology, biology of sex, and ecology, had, instead, a distinct genetic or evolutionary orientation. In institutions with affiliated medical schools, the biology department often provided preclinical training for medical students. But Lillie felt that this medical connection distorted biological research; it resulted in an overemphasis on problems of the individual at the expense of population issues. Lillie sought to rectify the problem by creating a biological institute that was separate from medicine. His opinions about the relationship between medicine and biology were similar to those of his mentor Whitman. When Whitman came to Chicago in 1891 he hoped to establish a biology program uninfluenced by the demands of medicine. Through his proposed institute, Lillie attempted to fulfill Whitman's goal.[27]

Although Lillie was quick to deny any affiliations with eugenics, arguing that "the propaganda along such lines is ill-advised and ill-timed," certainly the institute would lay the scientific foundation "on which social prophylaxis of the future must depend."[28] Lillie's reticence regarding eugenics was in keeping with the opinion of many biologists that the excesses of the eugenic campaign had brought it into scientific disrepute, but his reluctance may also be partly explained by the influential role embryology played in the department. The physiological, epigenetic orientation of developmental biology at Chicago contrasted sharply with the implicit preformationist and deterministic view of development held by geneticists such as Thomas Hunt Morgan. In Lillie's opinion, heredity did not completely define the organism; other factors influenced development. Life did not revolve around the germ plasm. The significance of embryology, Lillie noted, was that "it deals with a part of the life history which is determinative for all. How far it can be brought under control is a question of great social importance."[29]

Enacting a program toward the institute's establishment initially entailed additional appointments in cytology, genetics, and animal behavior. The department had been without a cytologist since the departure of Shosaburo Watase in 1900; yet the study of the cell, Lillie argued, was "basic to genetics and embryology."[30] In the field of genetics, Newman was the faculty member who approached the study of heredity along most traditional eugenic lines. He offered a popular introductory course on "Evolution, Genetics, and Eugenics" for over twenty years, changing its name to "Human Genetics" in the early thirties when the eugenic policies of Nazi Germany brought the dark side of eugenic reform to light. His own research centered on the study of human twins in an attempt to sort out the contributions of heredity and environment in development. Much to Newman's dismay, he found that in cases where hereditary differences were minimal, environment was a potent force in shaping human traits. The interaction between heredity and environment was much more complex than Newman had originally thought.

Sewall Wright was the other representative of genetics in zoology when Lillie submitted his 1930 proposal to Hutchins. Wright came to Chicago in 1926 to fill a position left vacant by the departure of Albert Bellamy for a post at UCLA. As a Child student, Bellamy was sympathetic to his mentor's physiological theories of inheritance, a fact that dominated discussions concerning Bellamy's replacement. Child was adamant that the department not hire someone trained in Morgan's school of "speculative predeterminism"; this left out H. J. Muller, a Morgan student under consideration.[32] Trained at Harvard under William Ernest Castle, Wright appealed to department members on a number of counts. First, Wright's interest in the physiological action of the gene seemed to fit within the departmental's focus on the physiology of development. Allee approved of Wright, for instance, because he believed him to be "one of the first of the geneticists to understand the bearing of Child's work on genetics." Child supported the appointment because he thought a person with Wright's mathematical skills would be a strong asset to the department, and Wright also came highly recommended by Child's good friend, Frances Sumner. Child was correct in his assessment of Wright. In fact, Wright's work in theoretical population genetics and his skill with statistical methods had a much greater impact on departmental members than his research in physiological genetics. Indeed, his shifting-balance theory of evolution helped orient and support the population focus of animal ecology at Chicago in the 1930s, a point to which I shall return later.[33]

Lillie was unclear as to how the additional appointment in genetics would fit within the scope of Newman's and Wright's research. He was more concerned that the person "possess insight in the social significance of [their] work," than he was with the person's specialty in the field.[34] What begins to emerge from Lillie's overall plan is the central importance that the control of population problems played in the institute's organization. Wright's own theoretical work, for example, focused attention on how population size and structure affected the course and direction of evolution. Although he questioned the scientific foundation of eugenics and was one of the more dispassionate or at least quiet members of the Chicago group with respect to the application of biology to human affairs, he did not object to the underlying principle of eugenics that biology could improve human society. Equally telling is the degree to which the study of population problems occupied the place of ecology in Lillie's institute.[35]

In the 1920s, ecology existed as a minor field in the department, although the situation improved somewhat with the hiring of Alfred E. Emerson in 1929. Emerson's appointment stemmed from concerns expressed by the university's president, Harry Pratt Judson, as far back as 1917 regarding the department's promotion, or lack thereof, of biology's ties to agriculture. The department took pride in its purification of biology from applied interests, and Lillie was not about to let Chicago become another agricultural experiment station, but he did indicate to Judson that there are "certain aspects of the control of insects and parasites that belong naturally in the department."[36] In 1926, the search for an entomologist began. Within the curriculum, the person was to teach field zoology and entomology and help alleviate the teaching loads of Moore and Allee. The new member would assist Moore with embryology and comparative anatomy and offer animal ecology alternate years so that Allee could take the summer quarter off to do research at the MBL.[37] Upon the recommendation of William Morton Wheeler, the offer went to George Salt, who had just completed a study on sugarcane borers at the Harvard Biological Laboratory in Soledad, Cuba. In the job description, the department emphasized the need for someone who would develop the "biological as distinguished from the economic aspects" of entomology. Salt was on a National Research Council fellowship and could not come until January 1929. Although he accepted the offer, he resigned before even coming to Chicago; he opted instead to stay at the Farnham House Laboratory in England where he began research on host-parasite interactions under the noted entomologist W. R. Thompson.[38] The position then went to Emerson, who had launched a career in the study of termites as a result of his associations with Wheeler at William Beebe's tropical laboratory in British Guiana. If the department wanted an entomologist uninterested in its applied side, Emerson was the right choice. He had a distaste for entomology's economic aspect and often complained of the intrusions on his time by pesticide firms that asked for advice regarding the effectiveness of various methods of termite control.[39] Consequently, the agricultural interests Judson hoped to cultivate lay fallow and forgotten.

Emerson did little to foster zoology's appeal to an agricultural clientele, but his interest in termite societies and their bearings on human social interactions strengthened the implicit agenda behind Lillie's institute. So did the work of Allee. Lillie pointed out to Hutchins the relevance of Allee's research on the biological basis of sociology for the institute's overall goals. But how did Emerson and Allee specifically see ecology contributing to Lillie's plans? Lillie divided the institute into five subject areas: sex research, genetics, embryology, ecology, and physics and radiation. Allee and Emerson saw their research cutting across four of these five areas. In the field of sex research, Allee was particularly interested in animal psychology, and he asked that an additional staff member be appointed whose research focused on nonmammalian behavior. This appointment would complement Karl Lashley's work on the neural basis of mammalian behavior in the Department of Psychology. Allee and Emerson both felt their research also touched on population problems in genetics. Their connections to the field of radiation studies were less direct.

Their main expertise was, of course, in the domain of ecology, and they advocated an approach rooted in population studies. In a memo to Lillie outlining ecology's relation to the Institute of Genetic Biology, Allee and Emerson noted that population problems had previously been studied through demographic and statistical techniques. Yet, the "physiological aspects of population densities," they argued, were an equally fruitful methodological venue. Population physiology differed from demographic population studies because the primary concern was not growth per se, but the effect of numbers on individual behavior. What influence did densities have, for example, on growth, survival, rate of learning, and reproduction? And what were the important integrating factors that united individual organisms into functional wholes? Such questions would be the mainstay of animal ecology research at Chicago throughout the 1930s and 1940s. In addition, Allee and Emerson hoped for the establishment of a permanent tropical ecology station situated in British Guiana. The rain-forest laboratory would enable researchers to make comparative ecological studies based on differing environments. Emerson, Allee, and K. P. Schmidt were all seasoned tropical biologists, and they shared a certain fondness for life in the jungle. But they also noted that the tropics furnished "the most available space for human expansion if the problems associated with such expansion can be solved." More than a romantic's vision of life in "primitive" nature permeated this proposal. For Allee, at least, the tropics offered a way of easing the tensions created by population growth, but the story of indigenous peoples and their interests was one that he naively forgot or cared not to hear.[40]

Apart from the faculty appointments in cytology, genetics, and animal behavior, Lillie's institutional plans included the construction of a new building which he estimated would cost approximately $1 million. The building would be twice the size of the Hull Zoological Laboratory and would be situated near the Whitman Laboratory of Experimental Zoology. This would bring zoology closer to the Departments of Physiology and Physiological Chemistry, facilitating interdepartmental interactions. In order to secure the large-scale funds necessary for such an enterprise, Lillie proposed the idea to the president of the Rockefeller Foundation, Max Mason, on 18 May 1931. Firm ties between the University of Chicago and the Rockefeller Foundation had already been well established. The initial founding of the university was dependent on a large endowment from John D. Rockefeller, and over the years the foundation continued to donate generous sums to the institution. From 1929 to 1934, the foundation contributed a capital grant of $30,000 per year to support research in the Division of Biological Sciences. Furthermore, the foundation indirectly supported sex research at Chicago through the NRC-CRPS making its total contribution to biological research at Chicago during the early 1930s over $50,000 per year. Only the California Institute of Technology rivaled Chicago for equivalent amounts of foundation support in the biological sciences during this period.[41]

Lillie's proposal for an Institute of Genetic Biology, requiring large expenditures, could not have come at a more inopportune time. The Rockefeller organization was undergoing a period of reorganization and consolidation. The Rockefeller boards, originally separate organizations consisting of the Rockefeller Foundation, the General Education Board, the Laura Spelman Rockefeller Memorial, and the International Educational Board, merged in 1928 into a single foundation with divisions in the natural sciences, the social sciences, the medical sciences, humanities, and medical education. Growth in the foundation's investments was seriously affected by the financial crisis of the Great Depression at a time when an even greater demand was placed on its funds. Although reluctant at first, in a conference with Lillie on 10 October 1933, Mason responded that he felt "*some* grant would probably be made," especially if matching funds were available from the university and a step-by-step plan could be worked out.[42]

Within the Natural Sciences (NS) division of the Rockefeller Foundation (RF), a new program in vital processes had been launched, one that corresponded closely to the goals of Lillie's institute.

Between the years 1929 and 1932, the direction of the NS division within the Rockefeller Foundation remained unclear. Questions regarding both the form and direction of support were largely left unanswered. With the appointment of Warren Weaver in 1932 as director of the division, however, a definite agenda emerged. Trained as a classical physicist, Weaver initially outlined a program for the NS division heavily slanted toward the physical sciences. Mason drastically cut Weaver's program but left one area intact. The NS division would promote projects in "vital processes"; this meant biological fields receptive to techniques and concepts derived from the physical sciences. Furthermore, the NS would give priority to fields that contributed to the study of psychobiology, which was the focus of the foundation's Medical Sciences (MS) division. Under this initiative, endocrinology, the biology of sex, experimental and chemical embryology, genetics, general physiology, biophysics, and biochemistry would receive support. Lillie could not have been more hopeful. The RF's directive in creating a "body of knowledge basic for the improvement of the physical, mental, and emotional status of man" was precisely the same organizational plan embodied in the Institute of Genetic Biology; even the subfields were almost the same. Funding the institute, Lillie argued, "would constitute the spearhead for the thrust which the RF plans to make." Yet there was a catch. Ecology was the one field in the institute excluded from foundation interests. To Weaver, ecology was glorified natural history; it could not profit from the physical sciences, nor did Weaver see it contributing to the study of human behavior. Lillie had to tailor his proposal to accommodate Weaver's demands. It was a move that had repercussions for Chicago ecology and foretold the bleak future that would continually plague attempts by ecologists in the 1930s and 1940s to secure Rockefeller funds.[43]

Beginning in 1934, the Rockefeller Foundation combined the NRC funding for sex research at Chicago with the general biological fund into a single grant of $50,000 per year. The reason was twofold. The NRC-CRPS decided to withdraw direct support of major projects such as Chicago's that were sufficiently stabilized; instead, they would focus on the small, less established projects and those lying in the field of psychobiology. Furthermore, as a member of the NRC committee, Lillie felt somewhat awkward appropriating such large sums of money for his own research. Direct support from the Rockefeller Foundation would eliminate any appearance of nepotism. In administering this new grant, however, Weaver reminded Lillie that money should not be given for research that fell outside the program interests of the foundation. This included, most notably, ecological research under Allee. As a consequence, Allee's Rockefeller grant (on the order of $3,000 per year) was terminated, and the money was used to support Sewall Wright's work in genetics.[44]

Lillie did manage to provide equivalent research funds for Allee from a separate university budget. Yet what appeared at first to be a simple transfer of funds blew up into a major incident. Allee not only lost Rockefeller support; he was also denied access to funds from the NRC-CRPS. When the Rockefeller Foundation combined the NRC money for sex research with Chicago's general biological fund in 1934, it did so with the understanding that the faculty would not submit individual applications for research that fell under the categories covered by both grants. Unable to get money from the general biological fund, Allee applied to the NRC-CRPS in March 1935 to support studies on the effect of hormones on dominance-subordinance hierarchies in birds. Although the committee awarded Allee $2,000, Robert M. Yerkes, the committee chairman, wrote to Allee that "for reasons which I cannot present by correspondence, I should advise you to decline the committee grant." Weaver was much incensed by Allee's application as it violated the informal agreement made between Lillie and the Rockefeller Foundation. Under the further persuasion of Lillie, Allee withdrew his application. He applied to the NRC committee the following year, only to get the same results. In contrast, Gladwyn Kingsley Noble of the American Museum of Natural History had no difficulty obtaining NRC-CRPS funds for similar research during the same period.[45]

Eager to accommodate Weaver's own interests, Lillie narrowed the scope of ecology in the institute's program to the study of experimental population problems. Even more significant, Emerson, who was a part of the staff in Lillie's 1930 proposal, was notably absent in the final proposal because his research on termite systematics was clearly outside the scope of "vital processes." In the formal proposal, submitted by Lillie to the Foundation on 1 February 1934, the overlap between the fields of inquiry in the Institute of Genetic Biology and Weaver's program is striking:

Lillie's institute	*Weaver's program*
Genetics: General genetics, human genetics, cytogenetics	Genetics: Cytology, genes, chromosomes
Embryology: General and experimental, human	Experimental and chemical embryology

Sex Research: Biology of sex, physiology of reproduction, psychology of sex	Biology of sex
Biochemistry with special reference to hormones	Biochemistry and biophysics
	Internal secretions: Hormones, enzymes
Means for physical analysis (biophysics), including radiation, and involving also the problem of environmental control	Radiation effects
Experimental population studies	Nutrition: Vitamins
	General physiology

Appealing to the larger interests of the foundation, Lillie emphasized the institute's concern with the "operative vital factors in evolution, with special reference to the progress of mankind." Pointing to the increased need for social planning during an era of social crisis, Lillie urged that sound scientific knowledge be secured in order to avoid "hasty and socially dangerous political action." The organization of the institute would draw together existing lines of biological research basic to the discernment of human population problems. Furthermore, in keeping with the overall plan of the Rockefeller Foundation, the institute was to be organized around projects that crossed disciplinary boundaries. Members of departments would receive concurrent appointments in the institute, and teaching would be carried out solely on the university budget.[46]

The $3 million Lillie requested would cover both endowments for new appointments and the construction of a building. Lillie envisioned the program developing in three stages. The first, for 1934–35, included construction of phase one of the building and the appointment of a cytologist. Stage 2 consisted of appointments in human genetics and in biochemistry, with a transfer of research in cytology and sex research to the new building. The final stage entailed the completion of the building, an appointment in the physiology of reproduction and experimental embryology, and staff provisions for physical analysis and environmental control. The ecology appointment in animal behavior included in Lillie's original plan was notably absent.

The RF decided to postpone a funding decision regarding the institute until fall 1934, after a study of the foundation's past operations and formulation of future plans was completed. Lillie was on the outside advisory panel that evaluated the NS program in "vital processes." He put in a plug for funding university research institutes and urged further support of the fields covered under Weaver's program.[47] Foundation officers and trustees thought otherwise. While supportive of Weaver's decision to develop biological fields on the borders of physics and chemistry, they were skeptical of the psychobiology concentration. The thrust of the MS division would remain in psychiatry and psychobiology; Weaver was to steer his program toward less clinically related fields in experimental biology. Weaver took this as a license to seed the nascent field of molecular biology, an area in which Chicago had limited expertise. Furthermore, the foundation, in a period of financial crisis, excluded institutional grants from consideration.[48] This new directive dealt a deathblow to Lillie's institute. Allied with the RF's interest in psychobiology and the "sciences of man" and dependent on a large institutional endowment, the Institute of Genetic Biology faded into the background. With retirement approaching in 1935, Lillie no longer had any incentive to pursue the project further. It remained a dream unfulfilled.

Although not a story of success, Lillie's attempt to build a center for the study of racial biology illuminates the institutional context from which a program of animal ecology materialized. Most striking is the degree to which many members of the department occupied common ground. Allee, Child, Emerson, Lillie, and Newman all shared a conviction that through biology, one could find solutions to the social ills threatening human civilization. The biologist was healer, capable of purging the disease and decadence that had festered in the bowels of human society and restoring balance and harmony to the social organism. This biological vision was by no means institutionally unique. Nor was their interest in the population as the organic entity upon which to direct their biological panaceas. But the particular amalgamation of physiological, organicist, and social discourse at Chicago lent credence to an animal ecology program, laden with social meaning, that emphasized the study of the population as a distinct entity and the integrating forces that unite populations into functional wholes.

In addition, the isolation of Chicago ecology from Rockefeller funds attests to the low profile ecology as a discipline had among foundation officers in the 1930s and the powerful influence Lillie had in garnering research money for embryology and sex research in the department. In 1931, Lillie was appointed dean of the Division of Biological Sciences at Chicago, and Child took over as departmental chair. When Child retired in 1934, Allee was "ambitious" for the chairmanship. He believed ecology had "not received quite the recognition that its followers deserved" and "felt under some obligation to work for [the] status" [of the profession]. Chairmanship of a department, he thought, was one way to increase the profession's visibility. Lillie dissuaded Allee from taking the position, although he gave mixed messages as to why. In 1930, beginning signs of paralysis appeared in Allee's lower limbs. The condition was caused by a benign spinal tumor, and Allee underwent a series of operations in 1930, 1933, and 1938, performed by the Chicago neurologist Percival Bailey, to remove the growth. After the first two operations, Allee walked with crutches, but the third resulted in complete paralysis in his lower limbs. From 1938 until his death in 1955, he was confined to a wheelchair, heavily dependent on the help of students and family to assist with the many problems that a parapalegic must face. Lillie cited Allee's medical condition as the main reason why Allee not take on the administrative load of chair, but he also "suggested vaguely something about the developmental field remaining the center of the department program." Lillie was indeed eager to see the embryological tradition at Chicago live on. Not surprisingly, Carl R. Moore, Lillie's graduate student and then colleague, assumed the chairmanship position.[49]

Lillie's retirement in 1935 changed the personal dynamics and power structure of the department. Even though developmental biology remained a departmental strength, it no longer held the authoritative position it once had under Lillie's reign. Consequently, the subordinate status of animal ecology at Chicago changed. Allee's close friendship and interaction with Emerson and the appointment of Thomas Park to the faculty in 1937, helped crystallize and initiate a core of ecological research centered on the population as the fundamental unit of ecological, evolutionary, and social change.[50]

6

The Integrity of the Group

The seeds of Chicago animal ecology were strewn across a developmental and physiological landscape, relatively isolated from the dominion of genetics and evolutionary theory. Yet the study of heredity by geneticists did not represent the outback of biological science, as some Chicago biologists such as Child believed, but the flourishing center. Mendelian genetics had grown into a major research frontier. And by the 1930s, the wall that previously separated experimental biologists working in the new area of genetics from field naturalists who observed evolution's tinkering in nature seemed to be cracking. Population geneticists—preeminently R. A. Fisher, J. B. S. Haldane, and Sewall Wright—were demonstrating mathematically that small selective advantages could produce evolutionary changes; they rarely dealt, however, with the naturalist's problems of speciation and the role of geographic and ecological factors in evolution. During the late 1930s and 1940s, a consensus among population geneticists and field naturalists concerning the precise mechanisms of the evolutionary process emerged, a period now referred to as the modern synthesis. Once held in biological contempt, natural selection, operating on small genetic variations, reappeared as the causal mechanism governing speciation and macroevolutionary change.[1]

Although Chicago ecology received its early sustenance from embryology, physiology, and animal behavior, newcomers in the 1920s, notably Sewall Wright and Alfred Emerson, brought into the rather inbred group at Chicago ideas from these evolutionary and hereditary domains. Lillie's Institute of Genetic Biology also provided a conceptual focus for thinking about population problems. In his studies of animal aggregations and communities, Allee considered the relationship between organism and environment to be the primary integrating force that united individuals into functional wholes. And these primitive associations laid the developmental foundations for the evolution of highly complex social organizations. This developmental view had been pervasive in the Department of Zoology at Chicago since its inception under Whitman in 1892. But the inclusion of Emerson and Wright in the department made Allee more sensitive to the fact that single-species aggregations were integrated, not just through shared behavioral responses to the environment, but through a shared gene pool—through heredity. In the 1930s, Allee began to reorient his theory of social evolution from a developmental framework of succession to a model more informed by Darwinian evolution. The 1930s are an important transitional period at Chicago, because one begins to see a research program in animal ecology coalesce that is focused on the population and is informed by different disciplinary domains. And it is important to not interpret this coalescence as the mere assimilation of ecology into the synthetic theory of evolution by means of natural selection. Indeed, the conceptual focus of Chicago animal ecology in the late 1930s emphasized the study of "population physiology" as much as the study of population selection. Synthesis can mean many things. In the case of Chicago ecology, it meant the synthesis of a physiological, developmental perspective with that of Darwinian evolution informed by Mendelian genetics.

Adding an Evolutionary Dimension

Compared to Ernst Mayr, Theodosius Dobzhansky, Julian Huxley, George Gaylord Simpson, and G. Ledyard Stebbins—the main "architects of the Synthesis"—Alfred Emerson's influence in spanning the interstices between ecology, systematics, and population genetics seems insignificant, especially since the synthesis has largely been defined by the architects themselves.[2] But the past historical focus on such luminaries as Mayr and Simpson has given a false sense of closure to a number of issues in evolutionary biology that continued to circulate in the literature, especially where evolutionary thought intersected with ecology and behavior. We still have a very inadequate picture of the rich texture and complexity of the synthesis that individuals in the "lower" echelons of evolutionary biology add. While most biologists by the 1940s believed natural selection to be the causative agent behind evolutionary change, still, the question of what level(s) selection operated on remained a highly contested and unresolved point. Emerson is an important figure in this regard. He not only brought an evolutionary dimension to animal ecology at Chicago, but he also provides a perspective on the discontinuities that pervaded ecological and evolutionary thought. A member of the National Academy of Sciences, president of the Ecological Society of America, Society for the Study of Evolution, and Society for Systematic Entomology, Emerson clearly achieved a certain degree of professional recognition among his peers. Although not an architect of the evolutionary synthesis by Mayr's definition, he was also not on the periphery.

Emerson began his professional career at Cornell, a leading center for the study of entomology under John Henry Comstock. Upon graduation in January 1918, he enrolled in Cornell's Ph.D. program in the Department of Rural Education, with a minor in psychology and ornithology. He had taught nature study classes for Anna Comstock as an undergraduate and expressed a certain delight in teaching biology to young children. World War I intruded on his plans, and he found himself in officer's training camp in the summer of 1918. Discharged the following December without seeing active combat, Emerson returned to civilian life to find an opportunity awaiting him that would decide the future direction of his academic career. His sister Gertrude managed to secure him a position as research assistant on an eight-month New York Zoological Society expedition to British Guiana under the direction of William Beebe.[3]

Beebe, curator of ornithology for the New York Zoological Society, established the Tropical Research Station near Kartabo, British Guiana, in 1916 at the request of Henry Fairfield Osborn, president of the American Museum of Natural History. Beebe was adamant that Kartabo not be solely a collecting site for museum specimens. When Emerson tried to arrange for his friend Karl Schmidt to accompany the 1919 expedition, Beebe expressed concern that Schmidt should not be coming just to build up Mary Dickerson's collection for the Department of Herpetology at the American Museum. Beebe was little respected among the professional biological community; he was more a popularizer of natural history than an academic research biologist. Yet at Kartabo he encouraged researchers to pursue problems in fields such as ecology, embryology, and evolution in addition to taxonomic work.[4] Although Beebe thus fostered an investigative environment at Kartabo, William Morton Wheeler, the renowned authority on social insects, proved decisive in guiding the direction of Emerson's research career. Emerson discussed his burgeoning interest in the study of termites, sparked by field observations made at the Tropical Research Station, with Wheeler in Kartabo during 1919. The following year, Emerson returned to British Guiana as a research associate, and during this visit spent a great deal of time with Wheeler in the field. In a letter to Schmidt, Emerson could hardly contain his enthusiasm. Wheeler was a "great zoologist," he exclaimed, and "I trailed after him to beat the band."[5] He returned to Cornell in January 1921, inspired to finish a doctorate in entomology by the summer. In the meantime, Wheeler recommended Emerson for an instructorship at the University of Pittsburgh to teach courses in entomology and ecology. In the fall of 1921, without a Ph.D. in hand, Emerson left with his wife Winnifred for Pittsburgh and launched a lifelong career in the study of the ecology, evolution, and systematics of termites. He soon became one of the world's leading authorities on termite systematics, and his personal termite collection, donated to the American Museum of Natural History upon his retirement from Chicago in 1962, contains 93 percent of the known species in the world (fig. 6.1). As a taxonomist, Emerson had much greater esteem for evolutionary theory than did Allee, and Emerson's interest in evolution eventually played a pivotal role in Allee's own understanding of population processes.[6]

In 1925, Emerson received his Ph.D. from Cornell for work on the phylogenetic origin of the soldier caste in the termite species *Nasutitermes cavifrons* (Holmgren). He continued to teach zoology at the University of Pittsburgh for another four years, whereupon he accepted a position as associate professor in the zoology department at Chicago. Allee and Emerson had, however, met previously in 1926, when Emerson taught at Chicago during the summer quarter. Walking along the Indiana dunes, Allee and Emerson established a close professional and personal relationship that would continue for the remainder of their lives. Writing to their mutual friend, Karl Schmidt, Emerson remarked that "Allee and I are aggreeable (*sic*) on almost everything from religion up to adaptation. He seems to think that animals in the artic (*sic*) are white because physiological conditions are similar rather than because of any protection they might get from the white background but never-the-less, we are not very far apart in our conceptions and it will doubtless do me a heap of good to hear the other side of the story for awhile."[7] Emerson was, of course, referring to a physiological as opposed to an evolutionary story of adaptation. The Chicago version of coloration in arctic animals would be couched in such terms as ecology, embryology, epigenesis, environment, and physiology in contrast to Emerson's account, which was dependent on a parallel discourse of evolution, genetics, predeterminism, heredity, and morphology. Although presented as polarized positions, Emerson came to embrace the physiological metaphors so pervasive at Chicago, while remaining committed to an evolutionary view. Nowhere is this more evident than in his reconsideration of caste production in termites.

Emerson's dissertation research arose from observations made in British Guiana of a soldier termite that emerged from the skin of a workerlike form. What was the phylogenetic significance of this,

Figure 6.1. Alfred Edwards Emerson. Courtesy of the University of Chicago Archives.

Emerson asked? Did this particular ontogeny indicate the evolutionary origin of the soldier from the worker caste? Although other termite specialists considered soldiers to be specialized workers, Emerson argued the reverse. He believed that the soldier originated phylogenetically from the first-form functional reproductive; the worker later developed in an evolutionary time scale from the soldier caste. But the most significant question was what biological processes governed caste development. Was this polymorphism genetically determined in the egg, or did the specialized termite castes develop in response to environmental factors such as food? Considerable debate on this point existed in the literature, although in the 1920s, the majority of termite researchers in the United States favored a hereditarian explanation. In 1926, Emerson argued that the castes were genetically distinct. Stimulated by discussions with his Pittsburgh colleague, geneticist Harold D. Fish, he suggested that a sterile caste such as the soldier could give rise to a worker caste in the following manner. Emerson first began with the assumption postulated by Darwin in *The Origin of Species* that neuter castes could only be explained on the basis of family selection. "In order that natural selection may have influenced this evolution," Emerson wrote, "it is necessary for us to assume that the parents that produce the best adapted sterile castes survive and pass on to their fertile progeny the power to produce the same type of sterile castes." Emerson then argued that a set of modifier genes could have arisen in the germ plasm of reproductive forms that would produce the sterile soldier caste. From the original soldier caste, Emerson maintained, a worker caste could emerge if another set of "determiners" evolved from the modifier genes governing the development of the soldier caste. This is what Emerson meant when he "referred to the phylogenetic development of one caste from another." In the evolution of sterile castes, Emerson thus envisioned an accumulation of "determiners" or sets of modifier genes, each set initiating the development of a particular caste. He did not, however, work out a precise Mendelian analysis to demonstrate how such polymorphisms would be inherited.[8]

The problem of caste determination was by no means resolved, and one of the first dissertations Emerson supervised at Chicago addressed caste formation in the genus *Reticulitermes*. C. B. Thompson, one of the leading proponents of the hereditarian origins of caste determination, argued that sterile and reproductive castes in termites could be distinguished at hatching by the size of the brain and reproductive organs. Thompson, however, had presented no measurements to substantiate her claims. Increasing the sample size of Thompson's original study threefold, Emerson's student, Laura Hare, undertook a statistical analysis of head-width to brain-width ratios in 150 individual termites at the first instar stage. Wright's mathematical abilities were an important asset to Hare's statistical work, as they were for many of Allee and Emerson's students. Indeed, many ecology students consulted with Wright on statistical procedures, and the supervision of this part of their dissertation was left in his capable hands.

Hare's measurements of termite head width–brain width ratios yielded a unimodal curve; no statistical differences in morphological characteristics between sterile and reproductive castes appeared until the third molt. Hence, different castes could not be discerned at the time of hatching. Hare was, however, cautious in her conclusions. She suggested that "special rate-controlling genes which are able to act only beyond a specific threshold" could inhibit or induce growth thereby generating differences in morphological characters among castes. But she was unwilling to state "whether the realization of this threshold is genetically determined or is influenced by extrinsic factors or whether both play a part."[9]

Hare's work partially undermined Emerson's previous conclusion that caste formation was dependent solely on heredity. The experimental investigations of Gordon B. Castle at the University of California further turned the tables in favor of an environmentalist interpretation of caste determination. When Castle isolated termite nymphs from functional reproductives, some nymphs developed into reproductive forms. If, however, a functional reproductive was present in the experimental group, no supplementary reproductives developed. Furthermore, when alcohol extracts obtained from the bodies of functional queens were fed to groups of isolated nymphs, the development of supplementary reproductives was delayed. Castle favored an inhibition theory of caste determination, first posed by A. L. Pickens, in which the reproductive caste was believed to exude a hormonelike substance that blocked the development of reproductive organs in the nymph stages. Every nymph had the potential of becoming a reproductive or soldier form at the time of hatching. The precise outcome depended on the caste composition within the colony. Thus, Castle emphasized the importance of "recognizing the integrity of the termite community," since the presence or absence of certain individuals would alter the developmental pattern of the colony as a whole.[10]

Emerson encouraged his students to explore the causal factors governing caste formation in more detail, and their experimental investigations supported Castle's findings. E. Morton Miller, for example, by grouping together various combinations of wing-padded nymphs, adult soldiers, and functional reproductives, found evidence of developmental plasticity among nymphs that was regulated by caste-specific inhibitors secreted by reproductives and soldiers.[11] Despite the abundance of data that supported a trophogenic (environmental) theory for caste differentiation in termites, Wheeler, at the time of his death in 1937, believed that caste determination in ants was blastogenic (of genetic origin). Robert E. Gregg, under Emerson's direction, sought to verify or disprove the blastogenic thesis of polymorphism in ants. He artificially created pure soldier and pure worker colonies by anaesthetizing wild nests and segregating the respective castes. Controls were maintained using colonies obtained directly from the field. Gregg's results indicated that the number of soldiers that developed in pure worker colonies was statistically greater than the number of soldiers that developed in either the soldier or control nests. Furthermore, the production of soldiers in pure soldier colonies was inhibited. Gregg could not completely rule out genetic origin since he did not transfer the queen from one colony into another, thereby maintaining genetic continuity. He felt, however, that the data strongly supported the presence of a "social hormone" in the soldier caste that inhibited soldier development.[12]

Two features stand out in Gregg's explanation of caste determination. First, Gregg argued that "the population as such cannot be neglected in an interpretation of the origin of polymorphism. The colony behaves as a unit and automatically approaches an equilibrium by adjusting the percentages of the castes if they become shifted from the 'normal' condition of the species." Second, Gregg suggested that caste characters were genetic, in that "each egg contains a complete set of all genes belonging to the species which necessarily includes those of all polymorphic characters." But the development of specific castes was determined by physiological thresholds created in both the internal and external environment of the organism that enabled specific sets of genes to become expressed.[13]

Gregg derived both arguments from a paper Emerson had given at a 1938 Cold Spring Harbor conference on animal and plant ecology in which Emerson revealed how far his ideas on caste differentiation had changed since his 1926 paper on the phylogenetic origin of the soldier caste. While "once he held the view that castes of termites were determined prior to hatching and thus . . . were genetically different," Emerson now thought otherwise. Each individual termite had the genetic potential to become any caste, just as every cell in the individual organism possessed a full chromosomal complex of genes. Caste development, Emerson argued, seems "to be caused by similar mechanisms to those influencing somatic development." Precisely what mechanisms did Emerson have in mind?[14]

Emerson, like his students Hare and Gregg, adhered to a physiological theory of gene action advanced by his colleague Wright. In a 1921 review of Child's book, *The Development and Origin of the Nervous System,* Wright had pointed to the importance of Child's physiological gradients in understanding the relationship between genetics and embryology.[15] This is why Allee thought Wright was one of the few geneticists to appreciate Child's work. Wright argued that the "specificity of gene action" was a "chemical specificity," presumably an enzyme, that guided "metabolic processes along particular channels" or gradients, in Child's terms. "Different genes," Wright maintained, were called into "play at different points, either through the local formation of their specific substrates for action, or by activation of a mutational nature."[16] Thus, the environmental complex created the conditions necessary for further growth and differentiation by activating certain genes and inhibiting others. But the environment was itself a product of past gene interactions. The gene, as Wright insisted, "did not represent in some way a particular morphological entity, but behaves rather as a modifier of the character of a physiological gradient." What Wright had done was to indicate how Child's gradient hypothesis could be genetically controlled. In hiring Wright, the zoology faculty at Chicago adopted a kindred spirit, one whose own physiological, organicist orientation struck a responsive chord. Development of a localized region, Wright insisted, depends on the "relational pattern of the organism as a whole." This was because the organism was a "highly self-regulatory system of reactions" governed by an "array of genes assembled in the course of evolution." Natural selection operated, not on individual genes per se, but on the organism as a self-contained reaction system.[17]

Emerson applied Wright's physiological theory of gene action to caste development by employing an organismic analogy. Termite societies are, of course, highly integrated biological systems, a fact which, along with the influence of his Chicago colleagues, proved of decisive importance in redirecting Emerson's thoughts along physiological lines. As Gregg noted in his explanation of caste determination, one could not neglect the behavior of the termite colony as a whole. If the termite colony was analogous to the organism as a self-regulated reaction system, the individual development in the nest was dependent on the relational complex of castes within the termite society just as the initiation or suppression of genes relied on intra- and interorganic environmental conditions. Indeed, the termite colony was the environmental complex within which individual termites develop and upon which natural selection acts.

As a student of social insects, and of termites in particular, Emerson was drawn to the view that the colony was the unit of selection. Explanation of the evolution of sterile worker castes on the basis of individual selection had always been problematic.[18] Because of the highly integrated nature of termite societies, Emerson adopted Wheeler's concept of the superorganism, in which the society was seen as an integrated, independent unit analogous to the individual organism. For

Emerson, the superorganism concept played a central role in explaining the integrating mechanisms of various levels of organization, subject to selection, and in comparing similarities, such as division of labor, social coordination, and social adaptation, between social insects and human societies.[19] Yet Emerson extended the superorganism concept beyond insect societies to include any integrated biological unit upon which natural selection could act. Thus, Emerson wrote that "in varying degrees, many of the dynamic factors which coordinate social populations also operate at simpler population levels such as the family unit, the sexual pair, and other types of organized animal aggregations. We seem to be forced to the conclusion that adaptive integration is largely due to the natural selection of various whole population units. The individual organism is not the only unit upon which natural selection acts."[20] Selection itself served as both the integrating factor within populations and as the common denominator that united all levels of biological organization.

Emerson's studies of termite societies led him to emphasize the primacy of the population as the unit of evolutionary change, an opinion he believed was consistent with Wright's theoretical work in population genetics.[21] In the early 1930s, shortly after the appearance of R. A. Fisher's *Genetical Theory of Natural Selection,* Wright advanced his shifting-balance theory of evolution. While Fisher based his quantitative analysis of gene frequency changes on the assumption of large panmictic populations in nature, Wright argued that population structure was more likely subdivided into a set of small semi-isolated groups or demes. Because of the small size of these groups, random drift has a greater effect on changes in gene frequency within each group than it would in a large panmictic population. Selection, moreover, is operating both within (intrademic selection) and between (interdemic selection) groups.[22]

Because Wright emphasized the importance of selection operating between groups, many of Wright's associates, including Emerson, Allee, and Dobzhansky, used Wright's work to underscore the importance of population-level selection. Wright's analysis stressed the significance of population structure in evolution in a way that Fisher's work did not. In Fisher's model, only individuals were selected for.[23] But because Wright divided the population into demes or groups that were themselves subject to selection, both ecologists and population geneticists assumed the shifting-balance theory to be an evolutionary model in which groups or populations were selected for. In the 1951 edition of *Genetics and the Origin of Species,* for example, Dobzhansky interpreted Wright's model in precisely this manner, especially when he applied the model to adaptive polymorphisms and race formation. Dobzhansky argued that balanced polymorphisms were advantageous to populations because they ensured genetic variability. Hence, "adaptively polymorphic populations should, in general, be more efficient in the exploitation of ecological opportunities of an environment than genetically uniform ones." In this argument, Dobzhansky imparted a certain organic identity to populations; populations had "organismic-like attributes" and the "rules governing the genetic structure of a population" were "distinct from those which govern the genetics of individuals."[24] This is why he referred to the physiology of the population, a term he borrowed from Chicago ecologists. His explanation of balanced polymorphisms relied on competition between populations, not individuals. Arthur J. Cain and Philip M. Sheppard took Dobzhansky to task because he assumed that polymorphisms were adaptive for the population without showing "whether the presence of polymorphism is really affecting the adaptedness of the populations concerned." In malaria-infested regions, for example, the gene for sickle-cell anemia persisted in heterozygous form because heterozygote individuals were resistant to malaria. Thus the polymorphism could be explained simply on the basis of individual selection.[25]

Emerson, like Dobzhansky, also used Wright's model to account for apparent population adaptations. Even in his systematics work, Emerson looked beyond individual variation to colony characteristics that could aid in unraveling the ecological and evolutionary factors responsible for speciation. On a series of expeditions to British Guiana, Emerson collected a number of termitophile species (guest beetles) from the nests of a single termite species *Nasutitermes guayanae* (Holmgren). The termitophiles were not, however, distributed randomly. Four species were found together in some colonies of *Nasutitermes,* while two different species were associated with other nests. Emerson suspected that the differences in termitophile species indicated the presence of two different populations of host termites, even though no ecological or morphological distinction in the host termites was observed. Upon further analysis, he found quantitative differences in soldier head length and bristle numbers that correlated highly with the termitophile distributions. Although he found them in the same locality, and they were morphologically and ecologically similar, Emerson concluded that the two populations of *Nasutitermes guayanae* were actually two sibling species, *Nasutitermes guayanae* and *Nasutitermes similis*. His understanding of the termite colony as an integrated system subject to selection thus directed his attention to the inclusion of termitophiles in a analysis of speciation problems.[26]

Since no apparent geographic or ecological barriers existed between *N. guayanae* and *N. similis,* the question of how these species originally diverged was problematic. Emerson suggested an evolutionary scenario akin to Wright's shifting-balance theory where close inbreeding and random drift within separate colonies could produce enough genetic differences in some physiological mechanism or reproductive behavior to "sufficiently isolate" the local population from the parent one, thereby creating a new species. Speciation of the accompanying termitophiles, he asserted, could result "if the germinal changes effected (*sic*) the chemical social mechanisms to a greater extent than the morphological characters."[27] On the basis of these observations, Emerson advanced a species definition in which the species is an "evolved, genetically distinctive, reproductively isolated, natural population." He argued that genetic distinction could be "morphological, physiological, or behavioristic" and that isolation could include any mechanism, so long as it "effectively prevent interbreeding with other populations."[28]

Emerson was active in trying to promote integration across the biological sciences in fields that touched on any aspect of speciation problems. As part of this aim, he helped found the Society for the Study of Speciation in 1940, which served as an informal network for U.S. biologists interested in "the dynamics of the origin of species." In addition, he acquainted ecologists with the latest developments in evolutionary theory by reviewing such influential texts as Dobzhansky's *Genetics and the Origin of Species,* Mayr's *Systematics and the Origin of Species,* and Huxley's *New Systematics* for the journal *Ecology.*[29] In these reviews, he often argued for his own species definition, but it was attacked by Mayr on two grounds. First, Mayr argued that two populations of a single species could be reproductively isolated by some geographic barrier, yet still potentially interbreed if the barrier was later removed. Second, Mayr thought the condition "genetically distinctive" was useless because "every individual has an overwhelming chance to be genetically distinct from every other member of his own species due to the almost infinite possible combination of alleles."[30] Mayr's criticism clarifies the way in which Emerson thought of the population in organicist terms. Emerson rarely referred to genetic variation among individuals. He accused Mayr of failing to "see the significance of the genetics of populations as opposed to the genetics of individuals."[31] When Emerson spoke of a population being genetically distinct, he had in mind variation among demes, in the same way that Dobzhansky discussed a Mendelian population with a "corporate genotype." Even though interdemic variation might be explained solely in terms of individual variation, drift, and migration, Emerson did not reduce interdemic selection to the individual level. Nor did he distinguish between altruism, which implies individual self-sacrifice for the good of the group, and mutual benefit, in which individual survival is increased by membership in a group without any energetic costs to the individuals concerned. To him, both were the result of population-level selection.

By arguing that the population was a distinct level of organization with its own inherent properties, population geneticists and ecologists in the 1930s and 1940s hoped to legitimize their respective disciplines.[32] Hence, even in instances like polymorphism, where a seeming population adaptation could be explained on the basis of individual variation and selection, many biologists still adopted population selection as the causal mechanism. To have done otherwise would have been to threaten the very existence of the population as a unit of study in a period when population ecology was just beginning to set down its roots as a subdiscipline within the soil of ecological science. This may be one reason why group selectionist arguments continued to flourish in the ecological literature long after many of the evolutionary issues of the synthesis were supposedly resolved. Only in the 1950s does one begin to see a shift among ecologists from "good of the species" arguments to individual selectionist accounts of adaptations such as clutch size. This trend gathered increased momentum after the publication of George C. Williams's book, *Adaptation and Natural Selection,* in 1966, in which Williams lobbied for a principle of parsimony in distinguishing between levels of selection. The principle of adaptation, Williams asserted "must be recognized at no higher a level of organization than is absolutely necessary."[33] By then, however, population ecology had become firmly entrenched.[34]

Emerson's studies of termite nests provides one further instance of how his own understanding of the population as a distinct entity subject to selection guided his evolutionary ecology research. In a paper published in 1938, Emerson argued that the study of termite nests was a study in the phylogeny of behavior. Termite nests were morphological expressions of underlying species-inherited patterns of behavior. Any possibility of inheritance of acquired characteristics was canceled, because the workers that built the nests were themselves sterile. Based upon similarities and differences found in nest patterns and correlations with morphological characters of caste, Emerson presented a phylogenetic arrangement of nest evolution in termites at the family and genera level. He also found cases of convergent evolution toward rain-shedding nest construction in different subfamilies of the Termitidae located in the tropical rainforest. Such adaptations in nest building, he argued, were instances of selection operating on the population as a whole.[35] The nest was a product of the group that could not be explained at the level of individual interactions. It was an outward expression, Emerson wrote, of "a social psychological pattern difficult to detect in the individual psychology of each worker."[36] As in his work with termitophiles, Emerson used nest-building patterns to inform his systematic work. On the basis of *Apicotermes'* nest-building patterns described by the termite taxonomist J. Desneux, he argued for the naming of a new species *Apicotermes gurgulifex*. Emerson felt the nests of *A. gurgulifex* differed significantly enough from the closely related *A. desneuxi* to classify it as an "ethospecies," even though no quantitative differences could be found among the morphological characters of the castes. His employment of ethological characters in delineating phylogenetic patterns paralleled similar work by continental ethologists such as Lorenz and Tinbergen on the evolution of instinctive behavior patterns in birds.[37]

Throughout his research career, Emerson remained committed to an evolutionary understanding of ecological and behavioral phenomena. In his analysis of nest construction, for example, he was primarily interested in the phylogenetic history of this behavior pattern which had, he insisted, a hereditary foundation. This was markedly different from the physiological and environmental focus that formed the cornerstone of Allee's own ecological and behavioral studies. In fact, Emerson saw his behavior research in the evolutionary tradition of Whitman, a behavioral approach that had been cast aside by Allee's mentor Shelford in the early formulation of animal ecology at Chicago.

Yet through their continued interchange, Emerson and Allee began to assimilate each other's ideas. Professionally nurtured and associated with institutions such as Cornell and the American Museum of Natural History, Emerson brought an evolutionary perspective that was decidedly lacking to ecological science at Chicago. By the late 1930s, Allee's writings began to pay much greater deference to natural selection as a causal mechanism governing the evolution of sociality. The developmental metaphor of succession that informed his earlier theory of sociality thus became partially replaced by a model that relied more heavily on competition between groups, natural selection, and heredity. And in abandoning succession, Allee also displaced the community as the focal point of study in favor of the population. During the 1930s, Emerson, in similar fashion, integrated the physiological focus of Chicago biology into his evolutionary perspective, as evidenced in his reformulation of caste determination and, more significantly, in his interpretation of the population as an organismic level of integration upon which natural selection acts.

The Biology of Growth

While population geneticists such as Sewall Wright explored the evolutionary consequences of population size and structure, animal ecologists were investigating experimentally the effects of environmental factors on population dynamics. This experimental work was often guided by a body of mathematical theory developed in the 1920s on the characteristics of population growth. Raymond Pearl's formulation of the logistic curve, in addition to the mathematical models of predator-prey interactions arrived at independently by the mathematical biologist Alfred J. Lotka and the Italian mathematician Vito Volterrra, stimulated ecologists to gather experimental evidence and field data in support or refutation of these mathematical claims. The British animal ecologist Charles Elton is an example of one pioneer figure who pursued field studies on the factors affecting population numbers.[38]

Today, one would generally characterize Allee's early animal aggregation research as an excursion into population ecology. In the 1930s, however, Allee had only a vague understanding of the population as a distinct level of ecological inquiry. For him, the aggregation was neither separate from nor independent of community-level processes. But, by the 1940s, the aggregation had taken on a meaning of its own. No longer an experimental subcomponent of community ecology, it was a fundamental unit of analysis: a population with its own properties such as birth rates, death rates, immigration, emigration, and genetic identity. The community, in turn, had its own qualities, with principles like the formation, stratification, and succession to study.

Lillie's Institute of Genetic Biology provided an organizational focus that helped initiate and clarify this shift in Allee's thought. Animal community ecology at Chicago emerged at the turn of the century in a nonhereditary framework. The community was an assemblage of species that came together as a consequence of similar physiological responses to environmental conditions. Changes in the community over time were epigenetic and progressive. By drawing attention to racial biology, however, Lillie highlighted the significance of studying animal groups associated by virtue of their hereditary relationships. Similarly, the research of both Emerson and Wright pointed to the importance of the genetic structure of populations. Consequently, Allee began to add an evolutionary/hereditary dimension to the aggregation that dissociated it from community questions. The transition from the community to the population in Allee's thought was also dependent on an increasing interest in problems of growth that brought him into the mainstream of population ecology research. In Allee's population studies, physiology continued to serve as the main guiding methodological framework, despite his inclusion of evolutionary mechanisms. And the main motivational factor continued to be the social significance of biology for humankind.

Allee's interest in growth focused on two closely related problems. At one level, the question was how biotic conditioning of the environment affected the growth rate of individuals. This research, which Allee started in the late 1920s, continued until his retirement from Chicago in 1950. In a certain sense, it was simply another assault on the nature of biotic interactions within the larger context of ecological succession. Early studies conducted in Allee's laboratory on the effect of biotic conditioning on growth indicated that isolated goldfish grew faster in water previously conditioned by other goldfish than in uncontaminated water. Later experiments demonstrated that the growth-promoting power of conditioned water resulted, in part, from the caloric value of regurgitated water.[39] The results were, however, complicated by the effects of trace concentrations of tin, copper, and aluminum ions present in the aquarium water.[40]

Although this research centered on the individual, the process of growth could also be examined from the standpoint of the population, an approach that had become increasingly popular through the mathematical forays of biologists such as Pearl demonstrating the general characteristics of the logistic growth curve. In both organisms and populations, growth is dependent on the multiplication of units, be they cells or individuals. During the 1920s, T. Brailsford Robertson in Australia and Pearl at Johns Hopkins illustrated the similarity between growth patterns in individuals and populations.[41] When plotted over time, the pattern took the form of an S-shaped curve and could be described using the general mathematical equation known as the logistic, first derived by the Belgian mathematician Pierre-François Verhulst in 1845 (see fig. 6.2).

Robertson derived his formula for growth, however, not from Verhulst's equation but from the equation used to describe autocatalytic chemical reactions. Pointing out the resemblance between autocatalytic reactions and growth processes, Robertson went beyond the mathematical similarities to suggest a common causal mechanism. In autocatalytic reactions, the product itself accelerates the ongoing

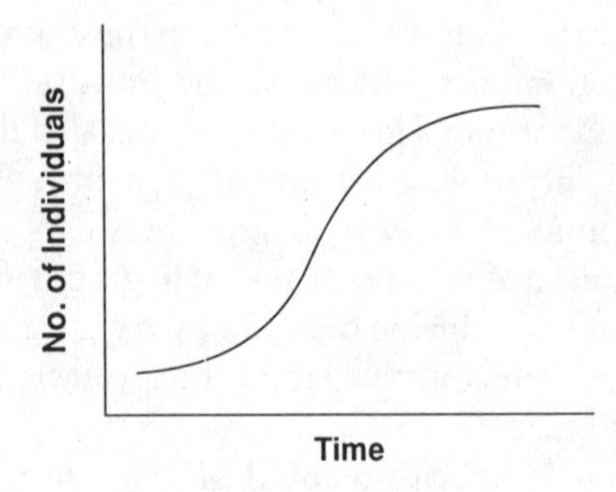

Figure 6.2. Generalized form of the logistic growth curve.

chemical process responsible for its production. Robertson hypothesized that growth too was an autocatalytic reaction whereby a growth-promoting substance that further accelerated the division rate was released from the nucleus during mitotic division.

Although his hypothesis was purely speculative at first, Robertson thought he found experimental proof for the autocatalytic nature of growth while studying infusorian populations of *Enchelys farcimen* and *Colpoda cucullus*. Two infusoria isolated into the same drop of culture medium reproduced at a rate more than double that exhibited by a single infusorian of the same species. Robertson attributed this phenomenon, which he called the allelocatalytic effect, to the release of a nuclear substance during division that accelerated the reproductive rate when present in the right concentration. He further suggested that at increased concentrations, this substance would inhibit growth, thus accounting for the decrease in reproductive rate that characterizes the later stages of logistic growth.[42]

Numerous investigators tried to reproduce Robertson's results but with little success.[43] The problem was particularly appealing to Allee, for here was another instance suggesting the beneficial value of animal aggregations, because the reproductive rate in Robertson's experiments was higher among grouped versus isolated animals. The allelocatalytic effect thus provided indirect confirmatory evidence for Allee's theory of sociality. Throughout the 1930s, the effect of population densities on reproduction and survival became a focal point of research in Allee's laboratory. But unlike the studies on individual growth, this research investigated processes such as density, survivorship, and fecundity that were intelligible only as population-level properties.

The allelocatalytic effect was the subject of a number of dissertations under Allee.[44] In the early 1930s, Willis H. Johnson was the first student, however, to provide evidence for a causal mechanism that explained Robertson's results. Earlier studies conducted in Allee's laboratory indicated that the effect depended on the relationship between the volume of culture medium and the number of animals present. But Johnson took the problem one step further. Postulating that food might be a causal factor, Johnson used a nonnutritive medium to control both the concentrations and species of food organisms (in this case, bacteria) present. Allelocatalysis in Johnson's experiments was not due to some mysterious *X*-substance, but was solely dependent on the densities of bacteria used as food. Concentrations of bacteria two to four times greater than the optimal food concentration resulted in higher rates of reproduction in the two-animal cultures. This was because the two-animal cultures could reduce the bacteria population to the optimum more quickly than single-animal cultures. When optimal concentrations of bacteria were used, no allelocatalytic effect was observed. Hence, Robertson's results could be "explained on the basis of the ratio existing between the infusorian and the bacterial population."[45]

Because of the difficulty involved in controlling the environmental medium in protozoan cultures, many investigators used nonaquatic organisms in their analysis of population density effects. Working with the fruit fly *Drosophila,* Pearl and his associates at Johns Hopkins found to their surprise that minimal population densities of *Drosophila* did not have the highest survivorship values. Instead, the optimal density for survivorship occurred in populations with intermediate densities. Populations with higher or lower densities had much higher mortality rates. Pearl's results on the relationship between population density and longevity were later confirmed by Friedrich S. Bodenheimer working at the University of Jerusalem.[46]

Whereas Pearl's results demonstrated a correlation between optimal population density and longevity, Thomas Park obtained results indicating the existence of an optimal relationship between density and fertility in the flour beetle *Tribolium confusum*. In 1928, Royal N. Chapman, an entomologist at the University of Minnesota and later Hawaii, conducted experiments with *Tribolium* demonstrating that an equilibrium averaging 43.97 beetles per gram of flour was reached even when the environmental size (quantity of flour) and initial population densities varied.[47] Interested in finding evidence for the beneficial effects of crowding, Allee reanalyzed Chapman's data and found that the highest rate of reproduction per capita occurred at an initial population density of four individuals. With densities above or below this level, the reproductive rate gradually declined.[48] The case was similar to Robertson's experiments with protozoan cultures.

Under the direction of Allee, Park experimentally verified the conclusion drawn from Chapman's data that the optimal population density with respect to reproduction occurred at intermediate densities. The result, Park suggested, was due to the interaction of two factors. As population densities increased, a greater portion of eggs was lost through cannibalism. Thus maximal population growth with respect to cannibalism occurred at low densities. But Park also found that copulation and recopulation stimulated egg production and led to higher fertility rates. Because crowded conditions facilitated increased copulation, the copulation factor favored maximal reproduction at higher population densities. Consequently, a population intermediate in size had the highest reproductive rate due to the interaction of these two opposing factors.[49]

Chapman expressed great concern over Park's *Tribolium* research. Part of the problem reflected a priority dispute, since Chapman was the person who introduced *Tribolium* as an experimental animal suited for population studies. Other leading investigators in the field, for example, Pearl, used *Drosophila* as their experimental organism. After receiving his Ph.D. from Chicago in 1932, Park went on a National Research Council fellowship to work with Pearl at Johns Hopkins. Chapman was somewhat distraught by these prospects, fearing that Park "under Pearl's influence" would "fail to do Chapman justice and that Pearl" would "walk off with the Tribolium project, leaving Chapman in the cold."[50] A rivalry already existed between Peal and Chapman, and Park's stay at Baltimore exacerbated Chapman's anxieties.

The second aspect of the problem related to Chapman's theory of environmental resistance. Drawing an analogy from Ohm's law, in which the electrical current is equal to the applied voltage divided by the resistance, Chapman postulated that the number of animals found in any population represents a balance between the maximal reproductive rate or biotic potential and the resistance of the environment to population increase. He reduced this concept to the equation

$$C = Bp/R,$$

where C = population density at equilibrium; Bp = biotic potential; and R = environmental resistance. In this equation the biotic potential or reproductive rate is presumed constant.[51] But in Park's experiments population density affected both egg laying and fertility. Chapman denied this possibility, maintaining that "all Tribolium under similar environmental conditions should have similar egg-laying rates."[52] According to Chapman, cannibalism, which fell under the general category of environmental resistance, was solely responsible for the observed equilibrium. Furthermore, Chapman was reluctant to accept the conclusion that maximal reproductive rates occurred at intermediate densities since this result implied that the biotic potential varied with changing densities. By separating reproduction from the physical and biotic factors of the environment in his equation, Chapman rejected the possibility that density could alter biotic potential.

Chapman's expressed concern over Park's interpretation of density effects was part of a sustained controversy in the ecological literature that began in the 1930s and continued well into the 1950s over the mechanisms of population regulation in nature. Charles Elton's influential text, *Animal Ecology,* published in 1927, brought attention to the problem of population oscillations. *Animal Ecology* was organized around the problem of animal numbers and the factors governing population fluctuations in nature. Elton maintained that for any given species, an optimal density existed, which was controlled to a large extent by predation.[53] The interest in population regulation during the 1930s also received an added impetus from debates among economic entomologists over the merits of biological control. Essentially, two contrasting positions emerged. The density-dependent faction, represented by such figures as A. J. Nicholson in Australia and Harry S. Smith in California argued that the population was in a state of balance or equilibrium, regulated by biotic factors such as predation or competition; in this sense, the population was its own self-regulating system. During the 1930s, F. S. Bodenheimer was perhaps the primary spokesperson for the alternative view. For him, populations were controlled by extrinsic factors such as climate, which operated independent of density.[54]

To a certain extent, the debate was related to the definition of the population itself. In general, those who placed particular importance on density-dependent factors saw the population as a distinct entity; self-regulation necessarily implies some degree of cohesiveness. As Sharon Kingsland has argued, the notion of the population as an ontological unit often stemmed from the mathematical theories used to describe population processes. By conferring biological reality on their own equations, mathematical biologists presumed a priori that the population was a physical entity obeying certain natural laws. Advocates for density-independent factors, however, rejected the existence of hypothetical population properties such as self-regulation and tended to emphasize instead the population as an aggregate of individual organisms. Criticizing the common view among density-dependent proponents that an optimum density for the population exists, and, furthermore, that such an optimum exists for "the good of the species," W. R. Thompson wrote:

> There is nothing in this which supports the idea of an inherent tendency toward optimum density. What the example really shows is that organisms multiply without regard to the species as a whole. . . . One should, of course, make an exception for animals living in organized societies. . . . In such species the society attempts to regulate its optimum density, though it often fails to do so. It is evident that such auto-regulation cannot be a property of mere agglomerates and that the idea of its existence, outside organized societies, has been reached *by transferring to the species as a whole the powers and tendencies inherent in the individual.*[55]

Interestingly, Thompson excluded species that exhibit social organization from his argument.

The issue of optimum population size in social species is tied directly to the ideas of Carr-Saunders presented at the beginning of chapter 5. Recall that Carr-Saunders argued that humans alone could regulate their densities by practicing birth control, and he provided a group selectionist scenario of how optimum population size would first arise. Many biologists in the 1930s and 1940s assumed that mechanisms of social organization in the vertebrates such as territoriality or dominance-subordination hierarchies enabled populations to regulate their size around some optimum mean. Thompson acquiesced on this point. In his criticism of density-dependent arguments for optimum population size, he admitted that social species might be the exception. Because social organization was believed to confer group advantage in competition with other less organized groups, natural selection would favor highly integrated and regulated animal societies. The explanation of optimum population size in animal societies was thus identical to the one advanced by Carr-Saunders in *The Population Problem*.

The widespread belief that social species could regulate their population size also has direct relevance when one considers the emerging definition of the population within ecological thought at Chicago. In 1937, Thomas Park returned to Chicago as an instructor in the zoology department. Park's appointment materialized because of the position left vacant by Child's retirement in 1934. Allee felt that he and the embryologist Paul Weiss could continue to represent Child's field of physiological zoology but only if a person was hired who had an interest in invertebrates and ecology. In addition, the physician, Percival Bailey, recommended that Allee no longer teach the field courses because of his deteriorating physical condition. Hence, an additional ecologist would relieve "Allee from his teaching work in ecology" and allow Allee "to center his activities on the physiological aspects of his program." Although Emerson might have picked up the ecology courses, he was "unwilling to shoulder the responsibility for carrying on the full ecological work of the department." Weiss lobbied for a cytologist instead, but the department voted in favor of Allee's wishes. The following year, he recommended that Park be hired for the job.[56]

Park brought with him an understanding of the population as a unique level of analysis. Having worked in Pearl's laboratory, Park was imbued with Pearl's philosophical approach to population biology, which emphasized the importance of considering groups as wholes rather than as a simple sum of individuals.[57] Emerson, as we have seen, regarded the population as a superorganism; it was a distinct entity, held together by integrating factors and subject to the forces of selection. By 1940, Allee had also come to understand his own animal aggregation research not as an aspect of individual or community ecology but as a distinct phase of population biology requiring its own explanatory principles.

In the *Social Life of Animals,* published in 1938, Allee presented two curves, which summarized the survival values and physiological effects of animal aggregations (fig. 6.3).[58] Curve *A* represents the density-dependent effects in logistic growth, where the detrimental effects of overcrowding become increasingly prevalent as the population size increases. The research by Robertson and Park, among others, on the relationship between population density and reproduction, however, indicated that often the smallest population size was not the most optimal. Hence, although the logistic curve illustrated the significance of density-dependent factors, Allee argued that inverse-density-dependent factors, which take a decreasing percentage of the individuals present as the population increases, also existed.[59] Curve *B* signifies the presence of inverse-density-dependent factors in which the optimal density is at an intermediate point. These beneficial effects of crowding were not apparent in the ordinary logistic growth curve, and a number of investigators, including W. Ludwig and C. Boost in Germany, Allee, H. T. Odum, and G. E. Hutchinson, discussed the mathematical implications of this social phenomenon.[60]

As discussed in Chapter 4, Allee interpreted the beneficial survival value of crowding at low population densities as evidence for nonconscious cooperation among animals. Sociality emerged from this protocooperative behavior. The animal aggregation helped ameliorate

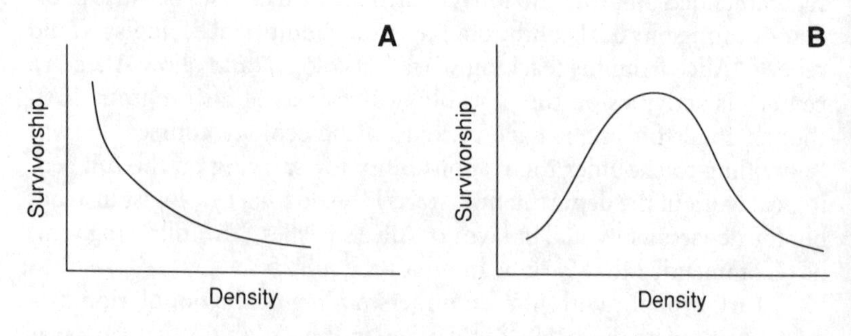

Figure 6.3. Curve *A* illustrates the harmful effects of overcrowding, showing the relationship between density and survivorship. Curve *B* illustrates the harmful effects of undercrowding as well as overcrowding. Here, the highest percent survivorship occurs at intermediate densities. Adapted from W. C. Allee, *The Social Life of Animals* (New York, 1938).

the individual's struggle with harsh environmental conditions and in the process made the physical and chemical surroundings more conducive to the survival of other more integrated and developed associations. In 1940, Allee offered an evolutionary argument for this theory of sociality that diverged from this earlier successional model, an argument that relied heavily on Emerson's superorganism concept and Wright's shifting-balance theory of evolution. Allee argued that the existence of optimal populations at intermediate densities resulted in "simple supra-individualistic groupings on which evolution [could] act to produce more complicated and better knit social systems."[61] If competition existed among groups, each possessing varying degrees of cooperation, then the less fit groups would be selected against. Note that cooperation was seen as the driving force in this evolutionary process and competition the effect. The group first arose from cooperation between individuals, which then resulted in competition at the group level. Because of the beneficial effects often associated with nonconscious cooperation, those groups that displayed greater cooperation among individual members would be continually selected for. For Allee, there was nothing novel in this interpretation. "All I am doing," he wrote, "is to recognize, along with Sewall Wright (1931, 1932), Dobzhansky (1937), A. E. Emerson (1939) and others, that populations are selected as well as individuals."[62]

What is distinctive about Allee's revised theory of sociality compared to his earlier successional model? By introducing natural selection into his ecological views, Allee incorporated a genetic component previously absent and thus gave definition to the aggregation as a population of related organisms, even though his understanding of genetics was at best incomplete. Like Emerson, Allee interpreted Wright's shifting-balance theory as one of selection operating on groups rather than on individuals; the population as a superorganism was subject to selection.[63] Furthermore, the inclusion of a hereditary component opened the door for nonprogressive theories of social evolution that were opportunistic in nature. Yet Allee did not follow this lead. He could not completely abandon the successional metaphor, where change takes place as a consequence of the historical unfolding of the social organism from the primitive to the more complex. According to Allee, the population, as an integrated unit upon which natural selection could act, was, by definition, social. Thanking Emerson for the "summarizing formula," Allee argued that "all groupings of individuals which are sufficiently integrated so that natural selection can act on them as units show at least the beginnings of sociality."[64] Hence, sociality began not with sex or the family but with the simplest aggregations that had group survival value. In his criticism of density-dependent explanations, Thompson argued that only animal societies had the power to regulate their own density. Allee evaded this criticism by arguing that the population was not a mere agglomeration of individuals but was actually a social phenomenon. He had injected his previous understanding of the community as a social organism into the population.

What united Allee, Emerson, and Park was the way they perceived the population as a distinct level of biological organization open to analysis and their insistent reliance on the population, not the individual, as the primary unit of selection. Although Park was much more reluctant than Allee or Emerson to invoke the superorganism analogy, for him, the population was still a distinct entity. And Park willingly admitted that the superorganism concept "does aid in focusing on the unity of the population by stressing the analogies or convergences between an individual organism and an organism group and by showing that this unity is in part a product of natural selection."[65] In all their writings, variation between individuals within a group or selection of individuals was rarely discussed. Thus Allee emphasized that "evolution, even the evolution of individual organisms is a process which takes place on the group level only."[66] For these Chicago ecologists, many ecological processes exhibited by populations could only be explained by reference to population selection, an evolutionary view which they argued was consistent with the writings of Theodosius Dobzhansky and Sewall Wright.

Building an Institutional Base

The assemblage of three animal ecologists within a single department gave the University of Chicago high visibility as a center for the study of animal ecology in the late 1930s and early 1940s (fig. 6.4). That visibility also extended to the professional sphere of editorial boards and governing appointments within scientific societies. Allee, Emerson, and Park all held the position of president of the Ecological Society of America (ESA) at some point in their careers.[67] In the 1940s, both Allee and Emerson were vice-presidents of the zoology section of the AAAS, and both were elected to the National Academy of Sciences. Their influence also extended to the journal *Ecology,* where Emerson and Park served as editors from 1932 to 1939 and 1939 to 1949, respectively. *Physiological Zoology,* under Allee's editorship, proved another important outlet for publications of Chicago students, which reflected the particular laboratory, physiology-based, behavioral orientation of animal ecology under Allee's direction. The "Ecology Group," an informal gathering of interested persons from the zoology department and other departments and institutions that met biweekly on Monday evenings throughout the 1930s and 1940s, also provided an important institutional core for animal ecology studies at Chicago. Here, graduate students, faculty, and speakers from local areas and afar were encouraged to discuss the latest developments and problems in the study of ecology as it was broadly conceived. At such a meeting in 1938, Park delivered a paper, "Concerning Ecological Principles," in which he attempted to delineate some of the major theoretical principles of ecology such as the "law of the minimum," succession, habitat selection, cooperation, competition, and logistic growth that could assemble and organize the accumulation of data on hand. His paper stimulated considerable interest among senior members in writing an animal ecology text on the basis of organizing principles.[68]

In the spring of 1939, cooperative venture on the book began with weekly Sunday meetings at which each chapter was meticulously read and criticized by all concerned. In addition to Allee, Emerson, and Park, two other individuals participated in this joint enterprise. Schmidt was a frequent visitor of the Ecology Group, and he served as the text's general editor. Orlando Park was the other contributor. A former student of Allee's and brother of Tom, Orlando was a faculty mem-

Figure 6.4. Department of Zoology faculty, circa 1945. Back row: Thomas Park, Herluf Strandskov, Carl Moore, Graham DuShane, Lincoln Domm; front row: W. C. Allee, Sewall Wright, Paul Weiss, Alfred Emerson, Karl Schmidt. Courtesy of the University of Chicago Archives.

ber at Northwestern University and his primary research interests focused on nocturnal ecology and seasonal periodicity of invertebrates in animal communities. The book would take ten years to complete, interrupted by the trials of the Second World War, the death of Allee's wife Marjorie in 1945, and an accident that struck Allee in 1946.

Allee's daily routine required that he wheel himself backward onto the platform of a home elevator that would take him to his room on the second floor. One February day, the elevator was in use in the attic, unbeknownst to Allee. He wheeled himself into an open elevator shaft, and landed head first on the concrete floor eight feet below. The fall resulted in a severe skull fracture, and while Allee recovered, he underwent a marked personality change. As his second wife Anne remarked, "the blazing temper which he had held under such iron discipline was given freer play than ever before." He "became a driver as well as a leader," berating his colleagues for failure to complete the promised chapters of the text on time. The project took on new force under his sometimes relentless rule.[69]

Published in 1949, *Principles of Animal Ecology* became known in ecology circles as the Great AEPPS (pronounced "apes") book, a punning acronym derived from the authors' last names. It represented the first synthetic animal ecology text structured around general ecological principles and organized according to increasing levels of integration, a pattern that many later ecology texts would loosely mimic. The book began with a history of ecology, followed by four main sections: analysis of the environment, populations, communities, and evolution. Allee wrote the section on the environment and the animal aggregations chapter within the population section. The majority of the population chapters were written by Tom Park, while Orlando summarized the body of research on community ecology. In the last section, Emerson discussed the significance of ecology for the study of evolution. The organismic, physiological focus pervades the work as does the importance of the population. As the authors indicated in the introduction:

> The reality and usefulness of the population as an ecological unit were apparent to us when we outlined the present book, and our subsequent work had reinforced our conviction of the importance of the principles that center on the population. We view the population system, whether intraspecies or interspecies, as a biological entity of fundamental importance. . . . The population is forged by strong bonds with autecology through the physiology and behavior of individuals; communities are composed of recognizable population elements; and evolutionary ecology depends directly upon population systems, since selection acts upon populations that evolve and become adapted to their environments to a more important degree than upon individuals. The study of populations as such, as operational systems, yields principles that clarify the nature of group interactions, interactions that do not exist at the level of the single organism, and that are too complex at the community level to be analyzed in a quantitative way.

Although *Principles of Animal Ecology* went through seven reprintings by May 1967, its impact on the future direction of animal ecology was relatively minor.[70] With a total of 837 pages, it was more of a reference work, encyclopedic in nature, to which researchers would refer for factual detail rather than for research ideas. It began in a period when the organismic, cooperationist ideal had been a mainstay of ecological discourse. In ten years, that discourse had shifted considerably away from cooperation to competition as a fundamental organizing principle and from an organicist to a cybernetics language of energy and information flow. The community as social organism was rapidly yielding to the ecosystem in which efficiency of information transfer was the strategy of the game. After the war, ecosystem ecology attached itself to the purse strings of the Atomic Energy Commission (AEC), quickly rising in prominence through the efforts of Eugene Odum at the University of Georgia and Stanley Auerbach and others at the Oak Ridge National Laboratory. Odum's *Fundamentals of Ecology*, published in 1953, became the manual of ecosystem ecology. Organized around Raymond Lindeman's trophic-dynamic concept and energy flow, it left the Great AEPPS book in its wake.[71]

In contrast to the future monetary success of ecosystem research, ten years of effort by Allee and Park to get funding for research in the field of population ecology was of little avail. The failed attempts of the NRC Committee on the Ecology of Animal Populations illuminates the shifting trends in ecological research that led to Chicago ecology's demise and the relative ineffectiveness of *Principles of Animal Ecology* in guiding the future direction of animal ecology research. In the summer of 1939, Frank Blair Hanson, associate director of the Rockefeller Foundation's Natural Sciences Division, approached Allee at Woods Hole with the idea of funding ecological research, particularly in the field of population studies. Hanson suggested an arrangement similar to the relationship that already existed between the Rockefeller Foundation and the NRC Committee for Research in the Problems of Sex. If a committee on population studies was established within the National Research Council, the foundation might consider making a grant to the committee. Hanson's idea was to fund a "small number of the very best men in this subject and to include basic research with animals from the lowest organisms on up and probably extending finally into the whole field of human social structure."[72]

Enthusiastic about the prospects for research funds, Allee wrote to Robert E. Coker, a member of the Executive Committee of the NRC Division of Biology and Agriculture. Coker was himself a former president of the ESA and had devoted his career to fisheries biology with particular emphasis on the ecology of freshwater copepods. Coker took the matter up with the chairman of the Division, Robert F. Griggs, a plant ecology researcher who had also been a past president of the ESA. Both Coker and Griggs encouraged Allee to pursue the matter further.

The NRC Committee on the Ecology of Animal Populations held its first meeting on 18 and 19 May 1941 in Washington, D.C. Members of the committee included Allee, Coker, Griggs, A. G. Huntsman, W. H. Johnson, T. Park, and L. J. Reed. Huntsman and Reed, apart from the NRC officers, were the only members who were not graduates of Chicago. Huntsman was a fisheries biologist from the University of Toronto, chosen to represent population studies in Canada. Reed was a statistician and close associate of Raymond Pearl at Johns Hopkins.[73]

The main goal of the first meeting was to provide a list of both important problems and investigators within the field of population ecology. At the suggestion of Reed, the committee decided not to enter the field of human population studies since this area was being funded by the Population Association of America. Instead, their focus would center on population studies of lower organisms, and from this research they would "arrive at the human aspect." Similarly, the work of population geneticists would not be considered since members felt they were "already receiving considerable aid." The committee then divided the field into three major categories: experimental populations, natural populations, and theoretical populations, with subcategories within each division. Rather than rank the problems in order of importance within each category, the committee combined laboratory and population studies and organized topics around the central theme of population physiology. The particular choice of problems thus centered on the functional relations of the population with its environment and on the factors that integrate populations into organic wholes. Important problems included quantitative studies on growth, population equilibrium, the "optimum yield" problem, the effect of abiotic factors and density on population growth and survival, competition and selection, population cycles, host-parasite relations, social facilitation, emigration and immigration, and range and territory effects in natural populations. Many of the problems reflected Chicago ecology's roots in experimental embryology, for the population was seen as an organismic unit that developed in space and time. Listed under each problem were those investigators who pursued research in the topic either from the natural or experimental standpoint. Both angles complemented one another, allowing the mutual verification of results. Theoretical aspects, which included "mathematical rationalizations, statistical methodology, and the social origins problem," occupied a separate domain.[74]

A formal proposal was submitted to the Rockefeller Foundation on 19 February 1942. Rather than serve as an advisory board to the Rockefeller Foundation, the committee felt it should have direct control of funds, which it would administer according to the recommendation of its members. The committee requested $25,000 per year for a period of three years. The proposal included a list of eleven specific research projects, eight of which were for direct support of committee members with a yearly estimated cost of $14,500. Of this money, $9,400 was earmarked for projects that directly supported the Chicago program of population physiology studies. If funds became available, the committee would canvass a larger number of researchers working in the field of population ecology for specific grant requests.[75]

Although the committee's major focus was the "attainment of an understanding concerning the ecology of populations," it also pointed out that important developments within other fields "have made information concerning the ecology of animal populations of key importance." These included (1) fieldwork in economic entomology, fisheries biology, and wildlife conservation where an understanding of population fluctuations was important; (2) a shift in attention among important geneticists "from individual heredity to phases of heredity among populations"; (3) "a developing interest in comparative studies among students of human sociology"; and (4) the discovery of nonconscious cooperation among animals. The latter three categories were all important components that guided ecological research at Chicago. As a further selling point, the proposal contained a set of appendixes written by various committee members outlining the significance of population studies for problems of aging, nutrition, human populations, and natural resources. Allee ended the proposal with a section delineating population ecology from the fields of individual and community ecology. "The population," Allee wrote, "exists within the community as a more or less definite aggregation that is sufficiently distinct to be readily recognized as an entity. The community itself is better understood as a loose organization, not of individuals, but of populations." By emphasizing the identity of the population, Allee hoped to underscore its importance as a unit of study worthy of research funds.[76]

In May 1942, Hanson wrote to Allee indicating the foundation had reservations about the project because of its policy to support "work in a certain field" which has "as a principal purpose the recruiting and training of specially able younger men."[77] With U.S. entry into the war, the majority of young men were in military service. Indeed, the foundation's copy of the proposal has the ages of each committee member penciled in next to his name. Only Park and Johnson were under forty; all the other members were in their late fifties or sixties and nearing retirement. Allee wrote back to Hanson arguing that there were still graduate students and investigators in early maturity who were not directly involved in the war effort. Furthermore, he felt mature investigators "should be aided in working to the approximate limits of their capacity while they are still able to do so" lest their "lives and their skills, especially in population research . . . be wasted." Allee's letter was to no avail. In September, Hanson wrote back rejecting the committee's request, suggesting that they resubmit a proposal after the war was over. In the interim, Hanson indicated the foundation would consider individual requests for grants-in-aid.[78]

The committee reactivated itself after the war with a slight reshuffling in membership. Reed resigned, and Harry S. Smith and G. Evelyn Hutchinson were appointed as new representatives.[79] Smith was an economic entomologist and director of the Agriculture Experiment Station at the University of California, Riverside, who had become well known for his work in biological pest control. Hutchinson, trained in zoology at Cambridge, had come to the United States in 1928 to occupy a position in the zoology department at Yale. Impressed by the work of the Russian biogeochemist V. I. Vernadksy, Hutchinson began in the 1930s to apply biogeochemical principles to the study of lakes, in particular Linsley Pond in North Branford, Connecticut. He was critical of what he termed the "bio-sociological bias of ecologists." Community ecologists had focused their attention on the "history of individual units and their interactions" without an appreciation for the dynamics and interactions of the physical environment with the biotic.[80] If the "community is an organism," wrote Hutchinson, "it should be possible to study the metabolism of that organism." Metabolism meant "transference of matter and energy," and Hutchinson sought to understand processes such as succession in thermodynamic terms. The potential of this approach reached its fullest expression in the work of Hutchinson's postdoctoral student Raymond Lindeman.[81]

Lindeman's classic 1942 paper "The Trophic-Dynamic Aspect of Ecology" utilized Alfred Tansley's concept of the ecosystem to consider the biotic community and its abiotic environment as a single functional unit. Building upon Elton's analysis of feeding relationships in communities, Lindeman transcended the particular nature of food and reduced it to a single principle—energy. The ecosystem could be understood as a system of energy flow from one trophic level to the next. Elton's pyramid of numbers could be explained at a more fundamental level; such a structural relationship was dictated by the second law of thermodynamics. Succession was now seen as a change in productivity of the ecosystem over time. Through Hutchinson, Lindeman had injected a quantitative and theoretical framework into ecology that opened a whole new domain of empirical investigation. Intimately tied to concepts derived from the physical sciences, ecosystem ecology flourished in the postwar era, funded by military agencies such as the AEC to trace the cycling of radioactive isotopes through the biosphere. In the process, the community as organism was transformed through the language of cybernetics into a self-regulated machine.[82]

Hutchinson saw mathematical theory offering similar quantitative insights into population ecology, as thermodynamics had for community research. The theoretical developments of V. Volterra, A. Lotka, and G. F. Gause were, he told Allee, population ecology's greatest strength. This body of "mathematical theory" . . . enabled "one to predict the consequences of hypotheses and then test the validity of the latter by comparing prediction with observation and experiment." In this respect, it was one "of the few branches of biology that has reached the degree of development achieved by 17th century preNewtonian physics."[83] Hutchinson was particularly interested in the application of the Lotka-Volterra equations of two-species interactions by the Russian ecologist G. F. Gause to experimental populations of *Paramecium*. Gause's experiments demonstrated that two closely related species in competition with one another partitioned different parts of the environment, enabling the species to coexist. In the late 1930s, Gause extended the "competitive exclusion principle," as it became known, from laboratory populations to a discussion of how such competitive interactions between populations in nature would yield overall patterns of community structure and stability. Hutchinson adopted Gause's emphasis on competitive relations to explain the abundance of phytoplankton species and their fluctuations over time in Linsley Pond.[84]

Although plant ecologists such as Frederic Clements had emphasized the importance of competition in succession, community ecologists working within an organicist framework interpreted competition as a cooperative force. By establishing patterns of dominance and subordination, competition created a functional division of labor that ensured greater efficiency and coordination in the social organism. This is why Clements and Shelford felt the need to coin the word "disoperation" in their 1939 text to denote harmful interactions between species. In contrast, ecologists such as Gause and Hutchinson, who were working in an explicitly nonorganicist mode, saw community and species distributional patterns arising simply from competitive interactions between closely related species. Like the ecosystem, competition became a major conceptual focus for ecologists in the postwar era.[85]

Both Orlando and Tom Park recognized the potential of ecosystem research and competition theory for restructuring the field of ecology. In *Principles of Animal Ecology,* Orlando devoted an entire chapter in the community section to a discussion of energy relationships and Lindeman's trophic-dynamic concept. His student, Stanley Auerbach, became a major figure in advancing systems ecology at Oak Ridge National Laboratory in the 1950s and 1960s.[86] Tom was likewise impressed by Lindeman's study. He was editor of the journal, *Ecology,* when Lindeman submitted his manuscript and eventually saw it through publication, even though it was rejected by two referees and looked upon with indifference by Allee. Furthermore, by the 1940s, Tom had devoted his laboratory system of *Tribolium* to detailed studies of competitive interactions in mixed-species systems. These experiments earned him sufficient respect among Rockefeller Foundation officers to attract $35,000 in funds between 1943 and 1959 and an RF National Sciences fellowship award to spend six months at Elton's Bureau of Animal Populations at Oxford University in 1948.[87]

The individual grants-in-aid benefitted Park, as it did Allee. Between 1943 and 1950, Allee received a modest $9,500 from the RF to support his aggregation research. But the renewed attempts by the NRC Committee on the Ecology of Animal Populations was unsuccessful in getting support on a national level. Griggs had become very impatient with Allee's attempts to sell population ecology by stressing its importance for understanding human social interactions. "It is fine for you to say that study of population problems," Griggs told Allee, "is the key to establishing the peace of the World. . . . If you could *prove* that, there ought to be loads of money to help you do the work. But as it stands now, there seem to be too many links in the chain of reasoning connecting research on animal population and the peace of the World."[88] Griggs was not alone in questioning the implications of Allee's research. While Griggs challenged the political leap Allee made from his biology, other ecologists expressed misgivings about the biological significance of mass protection. Hutchinson, for example, while not denying the evidence for the beneficial effects of crowding, argued that the importance of such density effects "in a general theory of population growth must, in the nature of things, be quantitatively insignificant."[89] In what was increasingly becoming a nature of competition, cooperation seemed little more than an anomaly.

With the death of Frank Blair Hanson, the NRC committee also lost its primary advocate at the Rockefeller Foundation. Warren Weaver as director of the Natural Sciences division had little respect for ecology; the withdrawal of RF support for Allee's research in the early 1930s attests to this. He felt that most ecologists worked "on so wide a front that they are spread very thin with essentially nothing but vague verbalization to show for their efforts." But as Allee discussed with Weaver the plight of population ecology in the late 1940s, Weaver developed a personal fondness for Allee. He admired the way Allee had "overcome his great physical handicap, and the good humor and energy with which he faces life."[90] Weaver also became interested in Allee's research on the social organization and behavior of the vertebrates, research that Allee began in the mid-1930s as an extension of his interests in the integrating factors that operated in the evolution of social life. Forced out by a mandatory retirement law, Allee left Chicago in 1950 to chair the Department of Zoology at the University of Florida. Weaver was instrumental in seeing that Allee receive $15,000 in RF money to support his studies on social organization at Florida from 1950 to 1955.[91] The irony is that Allee's work on dominance-subordination hierarchies ran counter to his portrayal of nature as a cooperative world. In exploring social organization among the vertebrates, Allee was legimating the importance of competition as a determining force in community structure, since dominance-subordination hierarchies were themselves established by individual-against-individual competition. This is perhaps why his research in social behavior had a much more noted impact than his aggregation work; it fitted more easily into the landscape of competition theory that came to dominate the field of population ecology in the 1950s and 1960s.[92]

This chapter has detailed the transformation of Chicago animal ecology from the community, heavily laden with developmental notions derived from experimental embryology, to the population, an intellectual axis closely aligned with developments in genetics and evolution. In defining the population as an organic entity, however, neither Allee nor Emerson nor Park completely broke from the physiological, developmental thought that had been so instrumental in the institutional germination of animal ecology at Chicago. By the late 1930s, these three individuals had established a core of ecological research outside the traditional hegemony of Chicago embryology, but they did not evade its influence.

The organicist language that abounds in *Principles of Animal Ecology* is indicative of the "biosociological" approach that characterized ecological research at Chicago and elsewhere during the interwar years. In viewing the population as an organism, both Allee and Emerson centered their analyses on the integrating forces that unite groups into functional wholes, be they animal aggregations, social insects, or human societies. Implicit in their analyses was the belief that through the study of animal societies, the biologist could bestow scientific wisdom and solutions gleaned from nature on the inherent problems confronting human society in an acutely troubled time. Yet, this extrapolation from the biological to the social realm came under grave suspicion as the organicist foundation of Emerson's naturalistic ethics, in particular, was seen to resonate too closely with totalitarian and fascist ideals in the war and postwar years. Indeed, Griggs's questioning of Allee's attempts to link animal population research with human social problems, suggests that a new postwar generation of ecologists was turning away from the scientific humanism that had been an important part of biology in Allee's generation. In fact, the popularity of ecosystem ecology in the postwar years signifies an important shift in the image of the ecologist and nature. The cybernetics language of the organism/machine embedded within ecosystem research reflects not the image of biologist as healer, but the image of biologist as environmental engineer, a manager of environmental systems. With the fears of strontium cycling through the atmosphere and the accumulation of pesticides in the food chain, ecosystem ecologists of the 1950s and 1960s looked to nature, not so much to learn about human society per se, but to monitor and solve environmental problems that human society had created.

While ecologists in the postwar years may have adopted a more secular attitude toward nature, scientific humanism has continued to pervade biological science, especially in the domain of animal behavior research, even after the Second World War. Although the racial hygiene movement of Nazi Germany brought into question the justification of political ideology on biological grounds, American biologists did not beat a hasty retreat from the human realm. With the increased threat of totalitarianism abroad, American biologists instead responded by erecting a biological defense of democracy that challenged the underlying principles of fascism. Nowhere is this more evident than in animal behavior research during the Second World War, to which Allee was a frequent contributor. But the organicist, cooperative renditions of nature that were the hallmark of Chicago ecology became increasingly difficult to sustain in the political climate of the cold war period in which group conflict and competition were seen as essential to a pluralistic, democratic society. A new ideological foundation for biological humanism was in order.

In the metamorphosis from community to population research and the inclusion of an evolutionary perspective, Allee found himself straddling two metaphors of a natural economy that did not mesh. Darwinian evolution did not so easily accommodate itself to the social, cooperationist orientation at the heart of his ecological ideas. He could not escape the fact that Darwin's metaphor was a struggle for existence between individuals in a harsh, competitive world. Park, in contrast, was uninterested in the underlying social ideology of Allee's ecology, and he both adjusted to and helped define an ecological view of nature as competitive. Allee and Emerson faced a more difficult task. They clung tenaciously to the cooperative metaphors so dear to their biological and social vision while adhering to a Darwinian framework posed by the modern synthesis. In the remaining two chapters, their science is explored within the context of World War II, in an attempt to detail the infusion of cultural and political concerns in their own understandings of the population as a biological and social entity.

From the Biological to the Social

The program of ecology advanced at Chicago, with its focus on the population, reached a threshold during the war years. Yet ecology at Chicago extended well beyond the bounds of internal scientific discourse. The population not only embodied a particular scientific approach, it also served as a window onto the social and ethical problems facing a depressed society immersed in a world war. The study of animal populations illuminated problems confronting human society and also facilitated scientific solutions to those problems. For Allee and Emerson in particular, ecology formed the basis of a scientific naturalism that functioned as both a legitimating force and a prescription for their own ethical and political views of a postwar society. In contrast, Park had little interest in the application of experimental analyses to human social problems; instead, his research focused almost entirely on laboratory studies of competition. But Park's inclusion in this study is important because it helps to establish a conceptual framework at Chicago—the population—that functioned equally well within the ecological profession and in the broader social discourse of American intellectuals struggling with questions of international peace and democratic order during the period of the Second World War.

In their social rhetoric, Allee and Emerson took certain underlying assumptions for granted. The individual, for example, played a subsidiary role to groups in the political process. In addition, emphasis on evolution by cooperation helped counteract the implication that war itself was somehow rooted in the Darwinian notions of "struggle for existence" and "survival of the fittest." Differences, however, are apparent in the natural economies that Allee and Emerson espoused. Although Emerson, like many postwar liberals, shared the hope in a cooperative world order, his understanding of cooperation was embedded in an organicist view of society that stressed the importance of social control in maintaining the functional efficiency and stability of society as a whole. Cooperation was important not as an end in itself but as a means to an end. This claim is examined by exploring Emerson's use of the superorganism concept to further his political views about democratic society and order within the context of social issues confronting biologists during the Second World War. Allee, on the other hand, was somewhat distanced from this mainstream component of scientific naturalism rooted in a natural economy that legitimized the existence of corporate capitalism and scientific management. As a member of the liberal pacifist movement during World War I and as an active member of the American Friends Service Committee during World War II, Allee embraced a social philosophy founded upon a politics of peace in which cooperation was itself the principle social goal. Through an analysis of Allee's research on vertebrate social organization and behavior and the conflicts that surfaced between his biology and politics, the underlying political distinctions between Allee's and Emerson's ideas come to light.

Science and Ethics

In his convocation address to the graduating class of 1940, Robert M. Hutchins, president of the University of Chicago, raised perhaps one of the most controversial and debated issues among Chicago faculty and students during Hutchins's reign. With the increase in totalitarian governments abroad and with the German occupation of France threatening America's neutrality, a great need had arisen to defend the principles of democracy. But "what shall we defend?" asked Hutchins. From Hutchins's perspective, Americans were standing on shifting sands. The century's preoccupation with naturalism had eroded the steadfast moral foundation embedded in metaphysics and theology. Under the lead of John Dewey, ethics had become a part of the scientific endeavor, where successful adjustment to the environment served as the ultimate criterion of right or wrong. Lamenting naturalism, Hutchins chastised the movement's intellectual leaders for "telling us in fact that nothing is true which cannot be subjected to experimental verification. In the whole realm of social thought there can therefore be nothing but opinion. . . . There is no difference between good and bad. . . . There are no morals; there are only the folkways. The test of action is success, and even success is a matter of opinion." The question was not so much what shall we defend, but how can we defend it? If ethics was relative, if moral choice was simply a matter of opinion, then how could Americans justify democracy as the best form of government. "Democracy," Hutchins wrote, "as a fighting faith can be only as strong as the convictions which support it. If these are gone, democracy becomes simply one of many ways of organizing society, and must be tested by its efficiency. To date democracy looks less efficient than dictatorship. Why should we fight for it?"[1]

Hutchins represented a widespread response among American intellectuals during the late 1930s to what historian Edward A. Purcell has called a crisis in democratic theory.[2] With the moral fabric of democracy threatened by fascism, some, like Hutchins, hoped to rectify the excesses of scientific naturalism by appealing to a renewed humanism in which a rationalist metaphysics served as the intellectual kernel of the higher education curriculum. Calling for educational reforms at the University of Chicago—reforms that reflected this humanist commitment—Hutchins accentuated the polarity and tensions among faculty members, students, and staff. As Mortimer Adler, a law professor at Chicago and one of Hutchins's greatest allies, recalled:

> just as Harper's Chicago reflected and formulated the "religion of science" which dominated American culture from the nineties to the thirties, so Hutchins' Chicago in the past ten years, has focused attention upon—more than that, has become the leading forum for—the crucial issue of our day: whether science is enough, theoretically or practically; whether a culture can be healthy, whether democracy can be defended, if theology and metaphysics, ethics and politics are either despised or, what is the same, degraded to topics about which laboratory scientists pontificate after they have won the Nobel Prize or are called to the Gifford Lectureship.[3]

Criticisms of scientific naturalism and science in general underscored the perception among scientists that their own profession was on the defensive. In their attempts to generate increased federal support for science after the First World War, many scientists emphasized the links between technological progress and pure-science research. During the 1930s, however, this Promethean image of science blurred into one of Pandora's box. For many, the economic plight of the Depression was due to an ever-expanding technology, one facilitated by the scientific enterprise. If science was implicated as a cause of the social crisis, then one possible response was to extend the boundaries of science into human affairs and, thereby, offer solutions to a society plagued by social problems and moral ills. Peter Kuznick has recently detailed the transformation of the American scientific community during the 1930s from an apathetic and distanced professional group within American society into a more socially aware and politically active force. Science could no longer legitimately be seen as functioning outside of society.[4]

Indeed, as Charles Friley, president of Iowa State College, reminded his academic colleagues in 1942: "the view that the sole function of science is the discovery and study of scientific facts and principles without regard to their social implications can no longer be maintained. Science cannot be divorced from ethics. Men of science can no longer stand aside from the social and political questions involved in the structure we have built from the materials provided by them, which structure their discoveries may be used to destroy."[5]

The war further challenged the amoral or, perhaps, immoral role of science in society. Surely the technological advances that flowed from the fount of scientific knowledge had increased the destructive powers of humankind. Raymond Fosdick, president of the Rockefeller Foundation, portrayed this negative image when he wrote in 1941 that "in spite of its claims and accomplishments science is to-day under sharp attack. The growing public realization that its powerful tools can be used for man's enslavement and destruction has given rise to bitter questions and charges; and we read to-day of a 'civilization betrayed by science' and of 'a degraded science that shirks the spiritual issues and hypnotizes its victims with its millions of gadgets.'"[6] But had science contributed anything to a positive social vision, to a greater understanding of the moral issues facing society in a period of crisis? Fosdick, unlike Hutchins, did not respond by resurrecting a theological or metaphysical framework. Rather, science as method provided the recourse and moral pillar for defending democracy and its associated values such as intellectual freedom and cooperation.

Edwin Grant Conklin's address as retiring president of the American Association for the Advancement of Science at Indianapolis on 27 December 1937 helped foster the appeal to science as the handmaiden of democracy and the ideological foe of totalitarianism. Conklin had attended the Blackpool meeting of the British Association for the Advancement of Science the previous year. There a proposal was made to draft a Magna Charta between the BAAS and the AAAS "proclaiming that freedom of research and of exchange is essential, that science seeks the common good of all mankind and that 'national science' is a contradiction in terms."[7] The AAAS address was the perfect forum for Conklin to bring issues already being discussed by his British counterparts to the attention of American colleagues. While Hutchins argued that science had eroded the moral fabric of civilization, Conklin saw otherwise. Following the lead of Hutchins's adversaries such as John Dewey and Sidney Hook, Conklin pointed out that the method of science was itself congruent with and served as the model for democratic faith.[8] In a series of lectures presented at Rice Institute in 1941, Conklin told his audience that cooperation and internationalism were an inherent part of the scientific enterprise. "Science," Conklin wrote, "furnishes the most striking instance in the modern world of human cooperation and genuine internationalism. It recognizes no boundaries of nations, races or creeds in its search for truth. Its progress depends upon real cooperation. . . . It is probably doing more than any other thing to break down the walls of prejudice that divide the world into hostile nations." An ethical code was ingrained in the spirit of science. Defending his profession, Conklin suggested that "if truthfulness, freedom, honor, humanitarianism, universal brotherhood are not ethical ideals, there may be some ground for claiming that the methods, results, and pursuits of science are unethical and are 'destroying western culture.'"[9]

Internationalism, cooperation, and intellectual freedom formed the ethical creed of science as professionals became more cognizant of their social responsibility during the war years. In the renewed search for a scientific basis of ethics, biology played an important role. During the late 1930s, a number of prominent American biologists including Edwin Grant Conklin, Samuel J. Holmes, and C. Judson Herrick published articles in which their science and, specifically, evolution provided the basis for a biological morality.[10] The similarity of their positions was evident in a symposium on science and ethics held by the Section on Historial and Philological Sciences of the AAAS in 1940. Under the direction of Chauncey D. Leake, biologists Holmes, Conklin, Herrick, Ralph W. Gerard, and others met with members from philosophy, sociology, economics, and mathematics to discuss whether human conduct developed according to some naturalistic principle. Formulating the principle in general terms, the participants concluded that "the probability of survival of a relationship between individual humans or groups of humans increases with the extent to which that relationship is mutually satisfying." This principle was, according to the group, merely a corollary of the general principle of Darwinian evolution whereby the "probability of survival of individual, or groups of, living things increases with the degree with which they harmoniously adjust themselves to each other and their environment."[11]

Across the Atlantic in England, an ongoing debate was taking place in the pages of *Nature* over an article published by embryologist C. H. Waddington on science and ethics. Waddington's position was similar to that arrived at by the participants in the AAAS symposium. Waddington saw the ethical principles of societies arising in the Freudian world of the superego; yet societies were not by nature static. Instead, they underwent developmental changes following a certain evolutionary trend. Hence what has survived is what is ethically good: "In the world as a whole, the real good cannot be other than that which has been effective, namely that which is exemplified in the course of evolution."[12]

These statements merely reaffirmed Dewey's pragmatist approach; if any moral absolutes followed, they existed only within the realm of the individual's own personal values. Successful adjustment to the environment constituted what was ethically good, but environments differed with differing time periods, cultures, and geographic locations. These examples, however, elucidate the sense in which scientists, and biologists in particular, were struggling to bring scientific authority into the social realm, to defend democracy on the basis of science, and thereby dodge the accusations of such critics as Hutchins to whom science appeared devoid of any moral or spiritual worth. Interestingly, while Hutchins called for a neo-Aristotlean revival at Chicago, his colleagues in biology including Allee, Emerson, and Gerard were actively advocating a naturalistic basis of ethics. Unlike the generalized statements of Leake and Waddington, however, their naturalistic ethics was linked to a specific research program that emphasized the population as a superorganism, the importance of group selection, and competition as both a cooperative and disoperative force. The focus returns once again to Chicago but this time within the context of the Second World War.

Levels of Integration

In September 1941, a conference entitled "Levels of Integration in Biological and Social Systems" was held at the University of Chicago to commemorate the university's fiftieth anniversary. Although the biologists and social scientists had originally planned separate symposia for the celebration, Thomas Park and Clyde Allee suggested to Robert Redfield, dean of the Division of Social Sciences, that the two programs be combined. The underlying purpose of the conference was to consider the functional relationships between parts and wholes, to examine the problem of integration within the unicellular organism through multicellular organisms, populations, and social aggregations culminating in human societies. The conference represented the first public collaboration between biologists and social scientists at the university. More important, however, as the title indicates, the symposium signified a common approach to the understanding of biological and social systems. Emphasis centered on the functional processes that facilitated biological and social integration.[13]

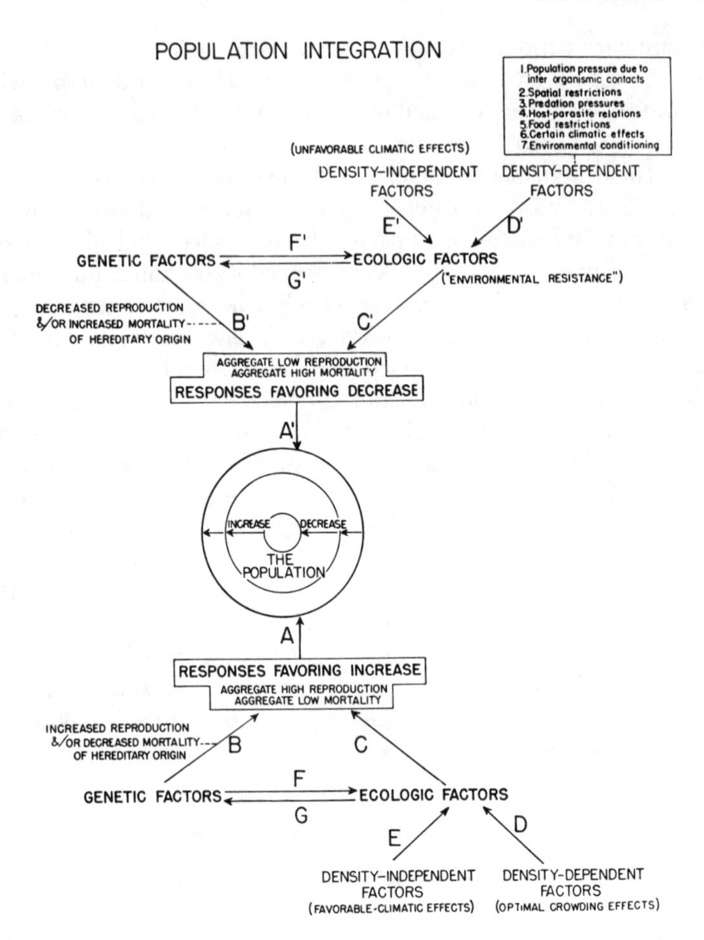

Figure 7.1. Thomas Park's diagrammatic representation of the factors governing population integration. From his "Integration in Infra-social Insect Populations," in *Levels of Integration in Biological and Social Systems,* ed. Robert Redfield (Lancaster, Pa., 1942), 123.

During the 1930s and early 1940s, the theme of organic functionalism echoed across the biological disciplines. Be it in embryology, physiology, or ecology, many biologists approached their subject with the assumption that the organism, population, or community was integrated into a functional whole; regulation of the internal parts such as cells, organs, individuals, or species ensured the stability, maintenance, and survival of the organismic unit.[14] Ecology at Chicago was no exception. As a distinct entity, the population had certain functional

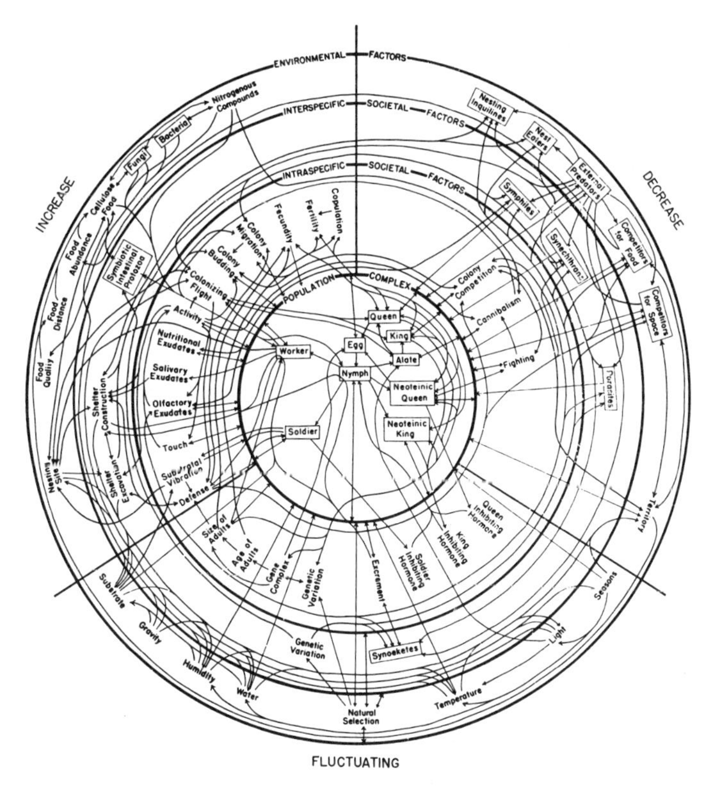

Figure 7.2. Alfred Emerson's view of the factoral interrelationships influencing a population of termites. From his "Populations of Social Insects," *Ecological Monographs* (1939), 288. Reprinted by permission of the Ecological Society of America.

properties that united its individual members into an organized whole. Working within this functionalist framework, Thomas Park, in his 1941 symposium paper, "Integration in Infra-social Insect Populations," presented a diagramatic representation (fig. 7.1) of the important factors that maintained population equilibrium over time. In an earlier paper published in 1939, Emerson presented a similar kaleidoscopic picture (figure 7.2) of the factoral interrelationships influencing a population of termites.[15] The ecologist, Emerson remarked, is interested in "the environmental effects upon population physiology and integration."[16]

For Emerson, termite societies represented the ideal model for understanding the mechanisms underlying population and social integration. As a taxonomist, Emerson consorted with many of the evolutionary theorists of the period; his ecological writings were, however, heavily laden with physiological metaphors. Writing to Allee, Emerson acknowledged his debt to "those who have been working on integration at the group or superorganismic level in contemporary systems" and saw himself differing "in important ways from those coming at the problem from an evolutionary background." Besides Allee, three figures loomed large in Emerson's discussion of biological and social integration: Charles Manning Child, Walter B. Cannon, and Ralph W. Gerard.[17]

Child left the urban environment of Chicago in 1937 for the northern California coast. Although he was not present at the 1941 conference, the numerous citations of his work indicate the integral role Child played in discussions of biological organization. The first three papers of the volume, written by Libbie Hyman, J. William Buchanan, and Ralph Gerard, built edifices around Child's physiological theory of dominance. Recall that for Child, the organism was fundamentally a behavior pattern; organismic form was not predetermined but emerged from the interaction of protoplasm and environment. Thus, activity determined structure. Integration of the organism depended on differentiation of the parts and physiological correlation of the parts into a unified whole. According to Child, two factors were important in biological integration: material or transportative correlation transferred material substances such as hormones between the parts while dynamic or transmissive correlation involved energy exchange. In higher organisms, the nervous system served as the chief conduit for transmissive correlation, transferring the excitation of living protoplasm to other parts. Child argued that in the transfer of materials or energy, differing rates of exchange are established which create physiological gradients. Dominance-subordination relationships originate from these gradient patterns; a hierarchy exists in which the most active part of the gradient, that is, where the rates of exchange are highest, becomes the dominant center regulating and controlling the flow of materials or energy to the subordinate parts. The most obvious gradient is found in the nervous system of the higher vertebrates where the centralized brain dominates the nervous system which in turn dominates and controls the entire body.[18]

Child's physiological theory of dominance was particularly useful for Emerson in viewing insect societies as superorganisms. In analyzing nest constructions of various termite families, Emerson noted that spherical, radial, and bilaterally symmetrical patterns occurred that seemed to correspond to symmetry patterns found in individual organisms. In the case of spherically symmetrical nests, a structural differentiation existed from the interior region where the queen was found to the exterior regions. The problem was how to account for this pattern. Since the same workers were responsible for construction of the nest interior and exterior, Emerson argued that the queen established an activity gradient that diminished as one moved farther away from the royal cell. Hence, "the queen constitutes a center of physiological dominance." The termite nest was an invaluable tool for understanding problems of physiological integration because it represented for Emerson an instance of "frozen behavior." The nest was merely a morphological byproduct of physiological activity. "The architectural nest," Emerson wrote, "is simply a more permanent expression of living activity gradients and its morphological attributes enable us to detect the results of the dynamic physiological events more readily." Resemblance of symmetry types represented a true analogy between the organism and superorganism.[19]

The analogy between the organism and superorganism went far beyond patterns of symmetry. The termite nest functioned to maintain a relatively constant and stable environment in the midst of externally fluctuating environmental conditions. In trying to assess the ecological functions of the termite nest, Emerson took measurements of internal nest fluctuations in temperature, humidity, and carbon dioxide. Although temperature varied with external environmental conditions, percent humidity within the nest remained fairly constant. In the terminology of the Harvard physiologist Walter B. Cannon, the nest functioned as a homeostatic mechanism, providing a stable internal environment for the termite society. From Emerson's perspective, homeostasis was the key link underlying all levels of organization. "One is forced to the conclusion," wrote Emerson, "that cooperative control of optimal environmental conditions for the organism composing the society is an important directional tendency in the evolution of the social insects, much as the cooperative control of optimal environmental conditions for the cells of multicellular organisms is an important directional tendency in the evolution of multicellular organisms."[20]

Emerson's unique contribution to the biological discourse on homeostasis and equilibrium theory during this period combined the concept of homeostasis with the principle of natural selection to arrive at a generalized theory of organic and social evolution. For Emerson, evolution always moved toward increased functional efficiency or homeostasis. He perceived a universal trend in evolution whereby the external environment of the cell or organism was internalized and brought under control. Once multicellular creatures had arisen, aggregations formed that helped ameliorate harsh environmental conditions. Ecological communities, in turn, developed which brought partial environmental control and stability to their individual members. But a much greater degree of integration came with the development of intraspecific populations. Here, cooperation among individuals resulted in a familial organization that reduced the severity of environmental fluctuations. And from the family, societies emerged that brought the environment under even greater control.

In this picture, Emerson felt he had rid holism of its mystical connotations. Natural selection operated to secure an ever-greater degree of functional efficiency; thus, selection accounted for the existence of internal adaptations that integrated the unit into a functional whole. Adaptations such as the evolution of mammary glands were to be understood not as ends in themselves but as factors that contributed to the increased homeostasis of the evolutionary unit—in this case, the population. The focal point of comparison between individual organisms, insect societies, and human societies was how integrating mechanisms led to increased homeostasis in each.

Into the Social Realm

In examining the writings of Walter B. Cannon and Lawrence J. Henderson on homeostasis and equilibrium theory in physiology during the 1930s, Stephen Cross and William Albury have suggested that the interest in organic stability among physiologists was not due to its value as a research guide; rather, it was a "form of participation in the culture of crisis of their time."[21] Cannon coined the word homeostasis in 1926 to describe the properties of self-control, regulation, and maintenance that ensured the stability of the "fluid matrix" or internal environment of the organism.[22] The sympathetic nervous system served as the primary mechanism by which the regulation of water, sugar, salts, and temperature were controlled and kept relatively constant in the midst of fluctuating environmental conditions. Homeostasis was not, however, a strictly physiological concept. In the epilogue to the *Wisdom of the Body*, published in 1932, Cannon drew an analogy between the body physiologic and the body politic. Responding to the economic instabilities that plagued society during the Great Depression, Cannon suggested the creation of regulative agencies "invested with power to preserve the constancy of the fluid matrix" of society. These agencies would regulate the processes of commerce (the fluid matrix of society) to ensure economic stability during periods of social crisis in the same way that the sympathetic nervous system controlled the fluid matrix within the individual organism.[23]

Cross and Albury penetrate beneath Cannon's immediate response to the problem of economic stability to suggest that much broader cultural concerns were embedded within organicist and functionalist models adopted in the biological and social sciences during the interwar years. "Many social commentators in the 1930s," Cross and Albury write, "went beyond the immediate problems of an ailing political economy to confront the apparent social instability of their time in broader and more speculative terms. They identified the crisis of civilization with the breakdown—as a result of rapid material change—of that spontaneity and naturalness of cooperation on which social stability seemed to depend."[24] At issue was the relationship between natural spontaneity and social control, between freedom and necessity. Indeed, during the 1930s, the appearance of fascism in Italy, Germany, and eventually Spain, and the crisis of capitalism and the widening powers of the state in America brought to fore one of the central questions of the period: What is the appropriate balance between freedom and necessity, between individual liberty and the powers of the collectivist state?[25]

The mobilization of the Popular Front in the late 1930s to combat the dark clouds of fascism in Europe exacerbated the issue of individual freedom versus state control and highlighted the need for a reaffirmation of democratic values.[26] Emerson's writings on homeostasis can only be fruitfully understood within this broad context. Like Cannon, Emerson used homeostasis within the realm of social discourse rather than as a research protocol. Despite his numerous writings on the subject, he conducted a minimal number of ecological investigations on the termite nest as a homeostatic mechanism. Further, in only two out of eighteen dissertations under his direction did the concept of homeostasis serve as a research model.[27] Writing a decade later than Cannon, Emerson was less concerned with economic instability than with the ethical issues surrounding fascism and increased militarism abroad. Yet parallel to other organicist models during this period, homeostasis and the superorganism concept helped resolve the conflict between the individual and the state and, from Emerson's perspective, legitimized the need for both individual freedom and social control. In homeostasis, Emerson felt he had arrived at a naturalistic basis of ethics that freed science from the accusations of critics like Hutchins who deemed naturalism responsible for the crisis facing democracy.[28]

Emerson was certainly conscious of the movement within the American scientific community to consider the broad social implications of science. The titles of his presidential addresses and university lectures given during the war years—for example, "Ecology, Evolution and Society" and "The Biological Basis of Social Cooperation"—

reflect this social concern. It is interesting that he always prefaced these popular writings with a statement about the negative implications of science. "Of what avail," asked Emerson, "are new lives if they face torture and starvation and new materials and inventions if they place power in the hands of ignorant, insane dictators. Is the 'law of the jungle' the contribution of biology to struggling mankind on the brink of the collapse of civilization? Does the scientist stop with the provision of greater power and more gadgets which may be used alike for common good or common disaster?"[29] These statements served as the perfect foil for Emerson to use in introducing his own biologically optimistic vision of society and in justifying the extension of biological principles into the ethical realm of human affairs.

In making the leap from biological theory to human society, Emerson first paid heed to the biology-culture distinction. Biological and cultural evolution are different, a difference which Emerson attributed to the communication through learned symbols found in human society. Cultural evolution occurred at a much faster rate, for change was not dependent on the slow processes of gene mutation and natural selection; instead, the use of language and symbols facilitated the rapid transmission of ideas within a generation. Hence, "the development of human social heredity through learned language symbols," Emerson argued, "is of such importance that this human attribute would seem to indicate the valid division line between the social and biological sciences."[30]

Yet the division between the biological and social was not so sharp. Although symbols and genes are fundamentally different with respect to their mechanisms, Emerson claimed that they are similar with respect to their function in both systems. Employing the evolutionary principles of convergence and analogy, Emerson maintained that useful comparisons could be made among organisms, insect societies, and human societies because each represented adaptations to similar environmental constraints. In both biological and cultural evolution, integrating mechanisms such as division of labor, gradients of control, communication, and cooperation developed that brought greater homeostatic control to the organism and society.

Of primary importance is the relationship between competition and cooperation in this evolutionary scheme. Emerson noted that with the trend toward increased homeostasis came an "evolutionary movement away from drastic conflict and eliminative competition toward more tolerable relationships."[31] Thus within the ecological community, cases of mutual harm or disoperation were rare, while the majority of relationships were characterized by unilateral benefit or mutual toleration. For example, competition between predator and prey led to greater efficiency in predation and better escape mechanisms on the part of the prey, signifying a case of unilateral benefit. Furthermore, because Emerson believed that selection could operate on various levels of organization, what appeared to be individual competition at one level may have been group homeostasis on another. A favorite example was the elimination of the puma by man in the Kaibab forest, a development that had disasterous consequences for the deer population. At the individual level, the conflict between an individual puma and an individual deer was obviously a ruthless struggle for existence. But viewed from the population level, the puma helped maintain and regulate the deer population, increasing the overall homeostasis of both. Depending on the level of integration, competition could have beneficial effects and thus be considered cooperative, or it could represent the harmful effects of disoperation. In 1945, at the Oriental Institute of Chicago, Allee, Emerson, Gerard, and Wright presented a series of papers on "Cooperation and Conflict in Nature." As the title indicates, the contrasting positions were not cooperation versus competition, but cooperation versus conflict.[32]

Instances of cooperation within the ecological community were, however, notably absent. According to Emerson, cooperation was, instead, characteristic of organismic and population levels of integration. Here again, one sees the primacy of the population in Chicago ecology. "Because of the survival or elimination of whole population systems," Emerson wrote, "cooperation between the individuals has evolved. . . . Just as the cell in the body functions for the benefit of the whole organism, so does the individual organism become subordinate to the population. It is in harmony with natural law to have an individual function for the benefit of other contemporary individuals and also for future generations."[33] At the heart of this statement lay Emerson's naturalistic ethics. For Emerson, homeostasis serves as the ultimate moral criterion of right or wrong. What was ethically right depended on how a given act enhanced or disrupted the stability or homeostasis of the functional whole. Just as the individual helped contribute to the stability of the population, so a person contributed to the general welfare of society. Cooperation was important not as "an end in itself" but as "the means to the end." The end, in this case, was greater homeostatic control.[34]

In building a social philosophy from the biological principles of homeostasis and evolution during the war years, Emerson continually emphasized certain themes. Just as there are important integrating mechanisms in nonhuman animals, ethics, according to Emerson, served as an integrating mechanism in human populations and groups. Drawing from biological generalizations found within the ecological community, Emerson saw the world moving toward increased integration, which required more mutual toleration. "Conflict," Emerson argued, "may be sublimated and controlled in the interests of peace, well-being and progress." As part of a generation that witnessed a lifetime of destructive conflict, Emerson longed for a period of social stability and harmonious relations. Still, Emerson distinguished between conflict, which implied violence and revolution, and more benign competition. Competition could be deemed beneficial, especially when viewed from the group level. As Emerson noted, "a certain degree of healthy competition between nations, races, denominations, classes, and institutions may speed progressive social evolution, but present destructive conflict is often deleterious to the world society and to the conflicting organizations."[35] Intergroup competition was a valuable process because it fostered greater social cohesion among members of a given group.

Emerson was fond of the phrase "united we stand, divided we fall" because it symbolized the centrality of the group in his social vision. By emphasizing group solidarity, Emerson shifted the struggle for existence away from intraspecies conflict to the struggle between a species and its environment. Here, cooperation was an important adaptive mechanism because it helped the population or group achieve greater control over external environmental conditions. The message for the human species in the midst of another world conflict was apparent. "The issue is clear," Emerson wrote three months after the bombing of Hiroshima and Nagasaki. "It is cooperation or vaporization. It is a struggle for existence by means of the cooperation of all mankind, or extinction through unnatural destructive competition between individuals, classes, races and nations already incorporated into a larger interdependent whole."[36] This emphasis on the group and environmental struggle was a common theme of cooperative theories of evolution dating back to Kropotkin.

Furthermore, the notion of the group, combined with Emerson's concept of dynamic homeostasis, helped resolve the conflict between the democratic ideal of individual freedom and the need for social control in ensuring a stable and efficient social order. For Emerson, individual behavior could only be evaluated by considering the degree to which it contributed to group homeostasis. Although individual freedom was deemed necessary for social progress, Emerson argued that "individual enterprise" was "not ethical if it rewards cleverness directed toward antisocial objectives." Social restraint was justified where individual freedom threatened the overall stability of the group. As Emerson noted: "Social pressures that inhibit or prevent such 'individual enterprise' are ethical if the result of freedom is a decrease in social health and social homeostasis. . . . Initiative and cleverness are not virtues in themselves. They may be deemed virtues only when they are directed toward individual and social progress." Hence, individual freedom was to be subordinate to the larger group. Both individual freedom and social control were necessary to ensure the maintenance of stability within society.[37]

By emphasizing group homeostasis and higher levels of integration, Emerson was stepping on dangerous ground. For his critics, such analogical reasoning bore a frightening resemblance to fascist theories emphasizing state supremacy. But Emerson felt otherwise. In both biological and social evolution, individual variation was necessary for progress to occur. Totalitarianism was flawed because it violated this fundamental precept of biological and social evolution. In his vice-presidential address to the zoology section of the AAAS in 1946, Emerson reminded his audience that "we have witnessed a gross misunderstanding of the operational factors of social evolution in totalitarian states that eliminate individuals with variant ideologies, thus attempting rather unsuccessfully to destroy the social variability upon which progressive evolution depends."[38] Emerson sought a compromise between individual freedom and state control in which increased homeostasis served as the ultimate moral criterion. In his eyes, a democratic society based on peaceful intergroup competition was the optimal form for progressive social evolution. It was a vision of democracy and of evolution in which the individual played a subsidiary role to groups in both the political and biological process.

Despite Emerson's hopes for a scientific basis of ethics, his naturalistic approach was essentially relativistic. There was no objective criterion that made it possible to judge between competing social systems. One commentator asked Emerson later in life, What criteria could you use "if alternative ways of maintaining . . . balance in society [exist], if there are alternative ways that are equally efficient?"[39] Certainly Emerson felt that his naturalistic ethics justified democracy, but his ideas could be easily misconstrued as a defense of totalitarianism, as his critics well realized. But Emerson was not the only Chicago faculty member attacked for his biological totalitarianism. Emerson's close friend, Ralph W. Gerard, was also criticized for expressing similar views.

Dangerous Analogies

Emerson's interest in organismic analogies and homeostasis was reinforced by his close associations with Ralph W. Gerard, a neurobiologist in the Department of Physiology at the University of Chicago. Gerard, trained under the tutelage of Anton Carlson and Ralph Lillie, received his doctorate in physiology from the University of Chicago in 1921. Like many of the best Chicago graduates in the biological sciences, he returned to Chicago as a faculty member in the physiology department in 1928.[40]

Gerard was himself a visible figure in the science-and-society movement during the late 1930s and early 1940s, publishing numerous articles on the social implication of science.[41] He was an active member of the Chicago branch of the American Association of Scientific Workers, an organization devoted to "promoting the interests of science and scientists and securing 'a wider application of science and the scientific method for the welfare of society.'"[42] And, like Emerson, he responded to this heightened social awareness by formulating a biologically based ethics, extrapolating biological theory into the social realm.

Gerard's naturalistic ethics and social vision closely paralleled Emerson's writings on the subject. Both started with the shared assumption that biological organization, from the cell to society, was characterized by increasing degrees of specialization and cooperative integration of parts into unified wholes. Emphasizing the levels of organization found in nature, Gerard adopted the term "org" to signify any system where interaction between the parts gave rise to a unit which persisted in time and space. Human society was, in Gerard's language, an "epiorganism," and as such shared many of the properties common to all levels of biological organization.

The important point about biological organization was not the structural details but the integrating mechanisms found within a given system. For a Chicago-trained zoologist, gradients were one of the most significant integrating mechanisms in both organisms and epiorganisms alike. Reflecting the influence of his teacher C. M. Child, Gerard emphasized that dominance-subordination patterns served as important powerful agents in uniting differentiated parts into an organized whole. The family was a case in point. According to Gerard, the parents, particularly the father, are dominant to the subordinate children. Eventually, as the children mature, the gradient pattern breaks down, and new interaction patterns are established, "as when the oldest son assumes the father role."[43]

Yet, an even more powerful and disclosing image can be found in Gerard's writings. Pointing to the specialization of sensory receptors and their dominant role in the highly developed nervous system, Gerard suggested a similar parallel in human society. "The social receptors," Gerard wrote, "which continually expand their sensitivity to the outer world are the scientists, and these units now point the direction of social evolution."[44] Clearly, Gerard saw scientists functioning as both leaders and experts in the management and control of human society. But they had hitherto shirked their social responsibilities, a trend that he hoped to rectify through his involvement with such organizations as the American Association of Scientific Workers. As Gerard commented elsewhere, "the scientist is not exercising, as he should, the gradient control which his role of receptor confers upon him."[45]

One of the overriding themes in these writings, as in those of Emerson, was the relationship between individual freedom and social control. Like Emerson, Gerard noted that with increasing integration came a greater degree of stability. But this control of fluctuating environmental conditions was achieved at a cost, for the org imposes certain constraints on the freedom of the individual parts to achieve this end. Yet Gerard did not see individual freedom threatened by this increased power of the group:

> The picture is commonly painted of ever more ant-like human societies with the individual reduced to a helpless slave of the group, lacking initiative, swathed in restrictions and altogether a sorry case. That social control will increase, I am certain; but that abject citizenry *must* result, I can not agree. . . . The org must modify the action of its units, but restrictions are balanced by new opportunities. . . . If we, in present are not "free" to go naked in the summer, neither were the Indians "free" to turn on steam heat in the winter. . . . We must learn to read and write; the peasant of yesterday had not this privilege. We may not kill, but radio, movies, books, even food packages with mystery serials, help us sublimate our aggressions.[46]

Through ever greater degrees of cooperation and self-sacrifice, the individual thus becomes part of a larger, more integrated, and more powerful whole. This evolutionary trend would not be complete, however, until human society had merged into a cooperative world order.

Gerard was quite aware that his thesis, like Emerson's, could be misconstrued as supporting totalitarianism. As an initial counter, he compared fascist regimes with the pathological breakdown of gradients of control such as occasionally occurs in the brain cortex. In addition, he argued that intense nationalism, as found in Nazi Germany, resulted in both isolation and overspecialization for conflict. Citing examples of the immense horns of the Irish elk and the oversized tusks of the saber-toothed tiger, Gerard claimed that such overspecialization was maladaptive and ultimately led to extinction. Furthermore, isolation and uniformity in Nazi Germany were the result of active coercion. But Gerard emphasized that "coercion of any sort, rather than persuasion and conditioning, is not an effective mechanism for obtaining a stable integration."[47]

Social control was to be achieved, not through force but through more democratic methods involving participation, group discussion, and effective leadership. Both Gerard and Emerson adopted a democratic approach to social control, one that historian William Graebner has labeled democratic social engineering. Their goal was to achieve a stable social order within the bounds of democratic process. Hoping to eliminate the dissociative powers of destructive conflict, they longed for a society in which individuals were "more cooperative, more consensus oriented," and "more group-conscious."[48]

Some biologists were, however, uneasy with what George Gaylord Simpson called the "aggregation" ethics of the Chicago school. Simpson, a paleontologist and celebrated figure of the modern synthesis, published a scathing attack on Gerard's "totalitarianist" biology, substituting in its place a naturalistic ethics based on his own views regarding individual selection. Simpson was troubled by both the biological and political ramifications of Chicago's ecological and evolutionary approach. For him, the individual was the fundamental unit of selection and the essential component of democracy. Rejecting Chicago's group selectionist views, Simpson saw the group as a mere amalgamation of individuals with no distinctive properties of its own. "The group," Simpson argued, "is a collectivity of individuals. It has no entity except as derived from the relationships of individuals. It does not evolve except as individuals prosper, and it is incapable of satisfaction but is modified and perpetuated by individual desires and attainments of satisfaction."[49] Emerson was, not surprisingly, critical of Simpson's ethical conclusions because of this individualist bias. Commenting on Simpson's *Meaning of Evolution,* Emerson remarked, "I found [it] somewhat frustrating because he did not arrive at some of the most important meanings due to his failure to grasp certain fairly well demonstrated principles. One principle was the unity of the group systems as entities in evolution."[50]

Group selection was not, however, the only problem that troubled Simpson. Simpson felt that the superorganism analogies of Emerson and Gerard neglected the fact that higher levels of integration could only be achieved by suppression of subordinate levels. Therefore, any biologist who employed the superorganism concept was forced into the position of accepting some form of totalitarianism. Commenting on Gerard's 1940 article, "Organism, Society and Science," Simpson remarked that the

> evolutionary analogy suggests . . . that the epiorganism will and should evolve in the direction of greater integration (i.e., less individual freedom and responsibility), and that its units (i.e., you and I) should become more specialized (with less scope for activity and change), more interdependent (less self-reliant), and more a part of the whole state (less individual). . . . Then the biologist finds himself face to face with the fact that this is a totalitarian idea.[51]

Simpson disagreed with the biological basis of Gerard's arguments against fascism and dismissed the whole organismic analogy as unsound. Neither party, however, questioned the propriety of basing ethics on biology.

For Simpson, evolutionary progress was not dependent on increasing levels of integration, as Emerson and Gerard advocated, but on greater individualization. "The integration that has been progressive in evolution," Simpson argued, "that has led to higher types of life and that has been 'good' biologically, or eugenic, has been integration of the individual."[52] For Simpson, social evolution was itself dependent on the evolution of greater individual capacities within the human species. Humans, Simpson argued, are the sole possessors of knowledge and also have the unique capacity for its transmission. This idea of "knowledge, together . . . with its spread and inheritance" formed the basis of Simpson's material ethic. For although such a statement appears amoral, it begins to take on an ethical significance if one considers "the promotion of knowledge" to be "essentially good." Science then takes on a moral flavor, for science stands apart as one of the most efficient and successful means of promoting and acquiring further knowledge. Simpson's ethics, however, was incomplete without the realization that humans have also acquired personal responsibility. Because of this, the *individual* alone is responsible for making right choices based on the acquired knowledge of humankind. Together, the ethics of knowledge and the ethics of responsibility prescribe the moral guidelines for human society; yet, these guidelines come first and foremost from the individual. Simpson's social ethics did not exist independent of individual rights.[53]

Critical of the aggregation ethics adopted by the Chicago school,

Simpson felt his views provided a more sound and rigorous support of democracy. But in this democratic vision, the individual was all important:

> The essence of democracy is belief in the importance and independence of the individual, and in the progress of society through the satisfactions of the individuals composing it. The essence of totalitarianism is belief in the unimportance of the individual and his subordination to the state, and in the progress of society as a thing in itself regardless of the satisfactions of the individuals in it. . . . I believe that it is our duty . . . to oppose the totalitarian fallacy and to maintain the true place of the individual in our social and in our biological philosophy.[54]

For Simpson, there was no balance between individual freedom and social control, no interplay between individual and group welfare. Instead, everything followed from the individual's inalienable rights.

While Simpson attacked Emerson's and Gerard's "totalitarian" premises, Alex Novikoff, a biologist at Brooklyn College from 1931 to 1945, later fired from the University of Vermont for his affiliations with the American Communist Party, challenged their analogical reasoning.[55] In his article, "The Concept of Integrative Levels and Biology," published in *Science* in March of 1945, Novikoff criticized the Chicago volume, *Levels of Integration,* particularly the papers of Emerson and Gerard. From a Marxist perspective, the biological and sociological constitute two distinct dialectical levels. Novikoff objected to Emerson's and Gerard's analyses because they violated this distinction by emphasizing the continuity between biological and social evolution. As Novikoff quite rightly perceived, "although Emerson acknowledges the distinction between biological and social sciences, . . . elsewhere, he maintains that 'the evolution of human social and ethical characteristics is governed by the same forces which have been directing organismic evolution through the ages.'" For Novikoff, human society constituted a distinct social level separate from the biological level of animal societies and required its own independent analysis.[56] T. C. Schneirla, a comparative psychologist at New York University and member of the Psychologists' League who supported numerous communist causes throughout the 1930s and 1940s, reiterated Novikoff's point in his criticism of Emerson and Gerard's deployment of the superorganism concept.[57]

Novikoff also objected to the notion that an organizing trend toward mutual toleration and cooperation characterized the entire evolutionary process, from molecules to human society. He feared that such ideas embodied an evolutionary fatalistic view of human society, leading ultimately to inaction. "Fortunately," he argued, "the United Nations are not guided by such fatalism; they are relying not on any 'trend,' but on their armed might, in order to defeat fascism and keep society on the road of progress."[58]

Emerson and Gerard published a rebuttal of Novikoff's criticisms in *Science* three months later. Their main response was to point out the validity and importance of convergent evolution in nature. Analogous functions exist between different organismic levels, they argued, and "inquiry into their origins and causes . . . leads, for example, to clearer formulation of evolutionary pressures as of how the environment operates through natural selection." They objected to Novikoff's refusal to entertain similarities between the sociological and biological levels, reminding their readers that Darwin's theory of biological evolution came, in part, from his reading of Malthus.[59]

The debate was not over, however, for Novikoff published another rejoinder.[60] But no new arguments emerged. The fundamental issue centered on the degree to which biological and cultural evolution could or should be viewed in *biological* terms. Novikoff worried that Emerson and Gerard were indirectly supporting fascism, "by making biological principles the guide for social thought and action."[61] Even though Emerson and Gerard believed their biological theories supported a cooperative world order rooted in democratic principles, Novikoff was right. Both Emerson and Gerard undermined the biology-culture distinction by emphasizing the continuity of the evolutionary process. And their understanding of this evolutionary process developed from biological, not sociological, theory.

Immersed in a world war, threatened by the efficiency and power of fascism, and challenged by humanistic critiques of science's moral worth, Emerson and Gerard hoped to bring scientific authority and order to a world gone awry. Their biological writings and theories were as much social statements about America during the Second World War as they were scientific opinions about issues in ecology or physiology. Indeed, for them, no distinction existed—biology was social. Certain themes synonymous with Chicago ecology were reflected in their social rhetoric: the population as a superorganism, the importance of group selection, and the significance of cooperation as an evolutionary force. Together these themes coalesced into a vision of a cooperative world order in which individual freedom and social control functioned to ensure a stable social order.[62] Although Emerson and Gerard hoped for a world free of war and destructive conflict, their main goal was to achieve a stable society. In this regard, cooperation was important as an integrating mechanism, an engineering device, which contributed to greater homeostatic control. Their natural economy, immersed in the politics of managerial capitalism, differed from Allee's in one important respect. Allee built his natural edifice around the politics of peace in which cooperation was itself the principle biological and social goal.

8

Building a Cooperative World

Emerson regarded dominance and subordination, principal mechanisms of integration in the individual organism and the body politic, as essential to the survival of insect and human societies. In adopting homeostasis or "self-regulation" as the evolutionary trend in biological organization, Emerson invoked a biological and political view of society in which dominance and control were important mechanisms for ensuring social stability. It was an assumption common in many organicist models of biological and social organization during the World War II period. Donna Haraway, writing on the history of twentieth-century primatology, has suggested that without the hierarchy, social order was "seen to break down into individualistic, unproductive competition."[1] With the threat of totalitarianism abroad, however, an inherent tension in the legitimation of dominance as a political and biological force surfaced during the Second World War, as evidenced by the criticisms leveled against Emerson and Gerard. Biologists struggled with the question of how one could combat the organicist ideology of fascism while still adhering to a biological view of the state. Emerson and Gerard opted to save organicism; they curtailed its darker side by insisting that fascism had been led astray in its failure to heed certain fundamental laws of biological and social progress such as the importance of individual variation and the maladaptiveness of overspecialization. Another solution, favored especially by a circle of biologists associated with Columbia University, which included George Gaylord Simpson and E. W. Sinnott, was to collapse organicism by placing the individual at the heart of evolutionary and social progress. Regardless of their differences, few individuals writing on the biology of democracy rejected naturalism as the basis for human ethics.

In the fields of animal behavior and social psychology, one could experiment with the consequences of group organization more directly than the mere analogical reasoning of such individuals as Emerson and Gerard allowed. During the decade surrounding the Second World

War, an extensive literature on the biological and psychological basis of aggression surfaced in America, a literature that, in general, emphasized the significance of learning and environment in the origins of aggressive behavior. Central to this research was how differences in authoritarian behavior led to differences in group structure and stability. Did an efficient group structure depend on the coercion of individual members by a dominant individual, or could some form of leadership, which relied on techniques of cooperation and participatory behavior, result in a stable social order that preserved the spontaneity of individual freedom deemed essential to democracy? Were animals by nature combative, and nations therefore doomed to perpetual war in the struggle for existence, or was aggressive behavior rooted in environmental causes under the biologist's control? These were a few of the questions that structured the animal behavior literature during the war years, a literature to which Allee and his students were frequent contributors.

From the mid-1930s to his death in 1955, Allee conducted research on social organization among the vertebrates, particularly the peck-order system in domestic fowl, to understand the physiological and psychological mechanisms at work in a relatively well-developed social group. Confined to a wheelchair in 1938, Allee was forced to limit his research solely to laboratory investigations; he could no longer endure the physical hardships of life in the field. Other members in the department, particularly Lincoln V. Domm, had extensive experience raising chickens for use in endocrinology research. The fact that much of the hormone physiology and psychology of these birds had been established was of decisive importance in Allee's choice of chickens as a research organism. They were also a familiar sight to one who grew up on a farm in the midwest. Yet Allee's preoccupation with dominance-subordination hierarchies, which are established by individual competition, seems contrary to the value he placed on the cooperative instincts found throughout the animal kingdom. Indeed, in his studies of vertebrate societies, Allee came face-to-face with more mainstream interpretations of sociality structured around sex and the family which he could not so easily evade. He did not deny the role of dominance as an organizing force in animal societies, but he did ameliorate its harshness by subscribing, like Emerson, to a group-selectionist view in which dominance was seen to have cooperative benefits. Unlike Emerson, however, Allee ultimately rejected dominance as an important integrating mechanism for achieving stability in human society. He envisioned a political order that was based first and foremost on the principle of cooperation. It was an international model based, not on dominance or social control, but on the politics of peace. In a certain sense, his research on dominance served as the perfect foil to portray his own political views.

Through his studies on the social organization of the vertebrates, Allee trained a number of students who had a significant impact on animal behavior research in America during the postwar era. Five of the first nine chairman of the Section on Animal Behavior and Sociobiology of the Ecological Society of America were Chicago graduates.[2] Similarly, when the *British Journal of Animal Behavior* opened its pages to American colleagues in 1958, six of the ten American members on the editorial board had Chicago affiliations.[3] Certain features of behavioral research unique to Chicago stand out, features wedded to the program of animal ecology nourished over the previous thirty years. Allee and his students sought causal explanations of vertebrate social behavior that were rooted in proximate mechanisms such as the organism's physiology or environment rather than ultimate mechanisms that focused on the organism's phylogenetic past. Although Allee did utilize natural selection to account for his theory of sociality and the significance of group organization, he was uninterested in the study of instinctive behavior for ascertaining phylogenetic histories. In this respect, his approach differed markedly from the nascent tradition of continental ethology being advanced by Konrad Lorenz and Niko Tinbergen, and the work of his contemporary G. K. Noble at the American Museum of Natural History in New York.[4] Furthermore, in his emphasis on the population as a unit of selection, Allee invoked a level of explanation centered on group structure in which social processes were not merely a consequence of individual behaviors. The group was not simply a byproduct of individual interactions that could be reduced in modern sociobiological terms to selfish genes, individual reproductive success, and reciprocal altruism but was a sociological unit integrated into a functional whole. Through an analysis of the animal behavior literature on dominance, leadership, and aggression during the Second World War, this chapter brings into focus the Chicago tradition of animal behavior research under Allee's leadership and the political context from which these studies emerged.

The Hierarchy of the Group

The study of flock organization, specifically of dominance-subordination hierarchies, received its original impetus from the claims of the Norwegian biologists T. Schjelderup-Ebbe in 1922 that despotism was the fundamental principle underlying the social organization of birds.

BW	pecks 9 :	W,	BY,	G,	RY,	B,	BG,	Y,	R,	GY.		
BR	pecks 8 :	W,	BY,	G,	RY,		BG,	Y,	R,		BW.	
GY	pecks 8 :	W,	BY,	G,	RY,	B,	BG,	Y,				BR.
R	pecks 7 :	W,	*BY*,	G,	RY,	B,	BG,			GY.		
Y	pecks 6 :	W,	BY,	G,	RY,		BG,		R.			
GB	pecks 5 :	W,	BY,	G,	RY,	B.						
B	pecks 4 :	W,		G,	RY,			Y.				
RY	pecks 3 :	W,	BY,	G.								
G	pecks 2 :	W,	*BY*.									
BY	pecks 2 :	W,				B.						
W	pecks 0.											

Figure 8.1. The dominance hierarchy in chickens, where BW is the alpha hen. From Ralph H. Masure and W. C. Allee, "The Social Order in Flocks of the Common Chicken and Pigeon," *Auk* 51 (1934):310. Reprinted by permission of the American Ornithologists' Union.

Schjelderup-Ebbe argued that flocks of birds were organized according to a rigid hierarchy in which the highest-ranking bird pecks all others but receives no pecks in exchange (fig. 8.1). The hierarchy was thus interpreted as a consequence of individual aggression that resulted in a stable organization.[5] By the late 1930s, spurred on by the funds and interests of the National Research Council for Research in Problems of Sex, the study of dominance, and especially sexual dominance, in animal societies had become commonplace.[6]

Although Allee had suggested as early as 1923 that the study of flock behavior might provide valuable insight into the nature and origins of social life, his first paper on social dominance and flock organization in birds did not appear until 1934. In this paper, Allee and a research associate, Ralph H. Masure, argued that the peck-order system described by Schjelderup-Ebbe was not the only type of dominance hierarchy. They observed, instead, that the common pigeon exhibits a less rigid type of hierarchy in which no individual has a definite peck right over another, an organization which they termed peck dominance (fig. 8.2).[7]

Allee's initial work on flock organization reflected the community orientation of his early animal aggregation studies. The study of flock organization was but one aspect of his investigations on the integrating factors important in community organization. In 1934, he commented on the ecological implications of this research:

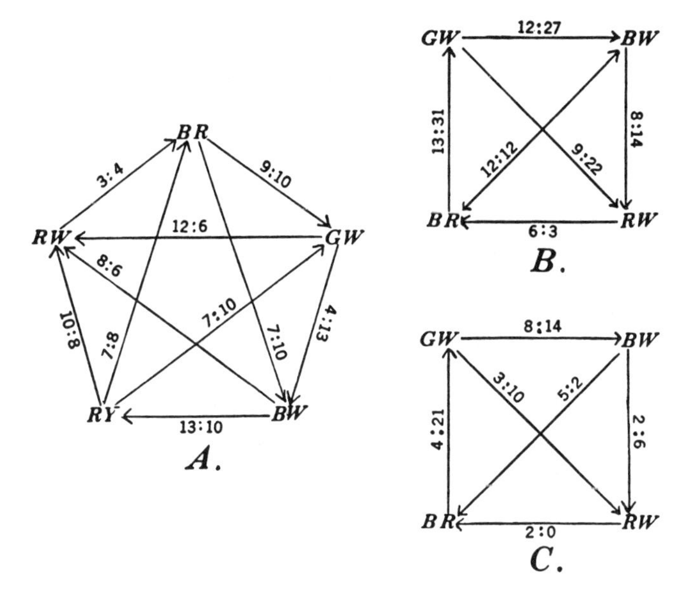

Figure 8.2. The peck-dominance organization in pigeons. Arrows indicate the relative status of individuals in relation to one another, but the organization is much more fluid than that of peck right. In *A*, for example, *RW* was observed to peck *BR* four times but received three pecks in exchange. From Ralph H. Masure and W. C. Allee, "The Social Order in Flocks of the Common Chicken and Pigeon," (see fig. 8.1). Reprinted by permission of the American Ornithologists' Union.

> Work such as given here indicates that there is probably in nature much more subtle intra-specific organization, particularly in flocking species, than that which is revealed by the well recorded territorial relationships. These are related to leadership and subordination, which, while difficult to study in the field, must be considered before we can have a final picture of the organization of an ecological community.[8]

Indeed, his analysis of flock behavior differed little from the physiological approach he adopted in studying animal aggregations. In 1935, he proposed injecting hormones into birds of known social rank to discover the physiological basis of the social hierarchy. This interest in hormones was not without precedent. Frank R. Lillie in zoology and F. C. Koch in the Department of Physiological Chemistry and Pharmacology were both actively engaged in work on sex hormones, research that was generously funded by the NRC Committee for Research in Problems of Sex and by the Rockefeller Foundation after 1934. Lincoln V. Domm, a former student of Lillie's and research associate in the Whitman Laboratories at Chicago, also investigated the hormonal effects on morphological and behavioral sex characteristics, especially among chickens.[9] Yet, this strongly entrenched research program under Lillie's direction became a financial hindrance rather than a benefit for Allee's proposed studies.

Allee applied to the NRC-CRPS in March 1935 to support studies on the effect of sex hormones on the social hierarchies of birds but was forced to withdraw his application. Under the terms set by Lillie, the Rockefeller Foundation, and the NRC-CRPS, individual faculty from Chicago were not allowed to submit applications directly to the Rockefeller Foundation or the NRC-CRPS for research in the biology of sex since this was already covered in the general biology fund. In 1936, Allee applied again to the NRC-CRPS to support the same research, but was once again turned down. Frustrated by this turn of events, he sent funding proposals to the NRC Committee on Research in Endocrinology, the Carnegie Institution, and even the Social Science Research Council.[10] In contrast, G. K. Noble at the American Museum of Natural History had little difficulty securing funds from the NRC-CRPS from 1935 to 1940 for similar experiments.[11]

Allee eventually secured a modest $4,000 from the NRC Committee for Research in Endocrinology for his hormonal research between 1937 and 1939. He and his students, including Hurst Shoemaker, Nicholas Collias, Catherine Lutherman, and Elizabeth Beeman, systematically explored the effects of various hormones on the social rank of individual hens. When a low-ranking individual in a flock of white leghorns was injected with testosterone proprionate over an extended period, the injected individual won a greater portion of staged contacts and rose in social status, eventually occupying the dominant position. Testosterone was presumed to increase the aggressive behavior of the subordinate individual, enabling it to secure a higher position within the social hierarchy. Allee expanded this study to include investigation of the effects of estrogen, epinephrine, and thyroxine on aggression, dominance, and submissive behavior.[12]

In these hormonal studies, Allee was trying to provide a physiological explanation for the existence of the social hierarchy. His work with testosterone seemed to verify a physiological basis for aggression. But Allee was also interested in determining to what extent learning or previous conditioning was responsible for dominance-subordination

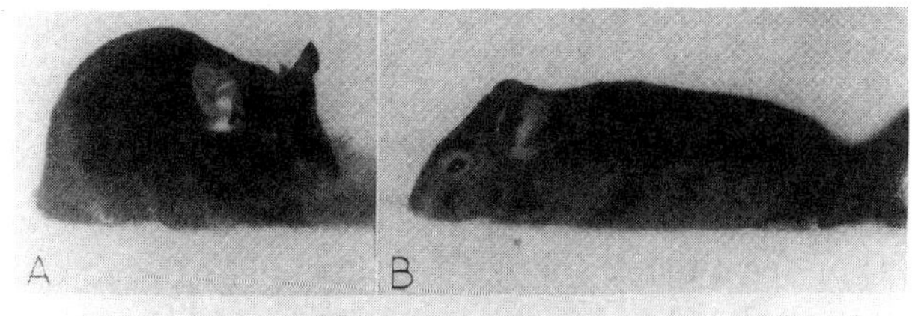

Figure 8.3. The "pacifist" and "aggressor" strains of mice bred at Jackson Memorial Laboratories. From J. P. Scott, "Genetic Differences in the Social Behavior of Inbred Strains of Mice, *Journal of Heredity* (1942), 13. Reprinted by permission of Oxford University Press.

behavior. Working with three inbred strains of domestic mice obtained from the Roscoe B. Jackson Memorial Laboratory at Bar Harbor, Maine, mice that exhibited genetic differences in aggressive behavior, Allee and his student, Benson Ginsburg, found that individual mice could be made less aggressive if exposed to repeated defeats in fights with other mice (fig. 8.3). Furthermore, they could also increase individual aggression by staging fights with more submissive mice, resulting in a continuous sequence of victories. It was, however, easier to "condition a socially superior mouse downward in the social scale by a series of defeats than it [was] to condition a socially inferior mouse upward."[13] Hence, although Allee and Ginsburg found hereditary differences with respect to aggressive behavior, aggression was also malleable, depending on past experience.

Research conducted by Allee and Beeman on the effect of thiamine (vitamin B_1) on aggression further emphasized the importance of conditioning in the emergence of dominance behavior. In the early 1940s, Allee was asked by Robert Griggs, Chairman of the National Research Council, to explore the relationship between degree of nutrition and morale in fighting; this research had obvious implications for American troops fighting in Europe.[14] The experiment indicated that thiamine deficiency did not affect aggressive behavior except for the indirect effects caused by physical weakness. Allee and Beeman concluded that "social status among . . . mice . . . seems to be influenced more strongly by so-called 'psychological' factors than by any other physiological consideration under our control."[15] If, as Allee suggested, past experience was more important than hereditary or physiological factors in governing dominance behavior, then education would be an effective tool in eliminating aggression in human society. Although Allee and his students acknowledged the importance of hereditary and physiological aspects of aggression, these were not the sole cause. Psychological conditioning always figured prominently in their explanations.

Viewed from the standpoint of the individual, dominance hierarchies benefited only those individuals at the top of the hierarchy by conferring on them the privilege to obtain food and mates. Although Allee recognized that high social rank in the peck order contributed to an individual's reproductive success, his central concern was whether dominance-subordination organizations provided any group advantage. His main interest was in how organizational patterns affected the behavior of the individuals within the group, a group which was itself a product of individual interactions. But the group was more than just a product, it was itself a guiding influence. This idea was central to Chicago embryology and animal ecology. In Child's regeneration studies, for example, the organizational pattern laid down in the interaction between protoplasm and environment determined the future development of individual cells. If one experimentally altered the organism's polarity, one could induce individual cells in *Planaria* to dedifferentiate and take on new functional behaviors. Child perceived the organismic whole as the environment to which individual cells responded. Organization thus influenced all present and future behaviors and was at the same time a product of past interactions between the cells and their intraorganic environment.

In the summer of 1942, Allee and his student, Alphaeus Guhl, designed an experiment that elucidates the interactionist, sociological approach that characterized Chicago animal behavior research.[16] Allee and Guhl noted that it was relatively easy to investigate "the possible correlation between individual success and individual social status."[17] Aggressive individuals who rose in social status generally left more offspring than those low in social rank. But this was not the focus of their research. They wanted to know instead how the environmental context of different group organizations affected individual behaviors and success. To investigate this problem, they designed three different hen groupings: three control flocks that had established a peck order, one experimental flock in which a single hen was removed and replaced daily with a hen that had been in isolation, and a number of isolated hens. They then compared such factors as food consumption, body weight, and egg laying among individuals across the different groupings. Their results indicated that hens in the socially unstable flock consumed less food, lost more weight, laid fewer eggs, and tended toward smaller comb size than hens in flocks with an established dominance hierarchy. Differences in behavior, they concluded, were not due solely to individual variations but were caused by effects associated with group membership. Social processes were not seen as a consequence of individual properties but were instead determined by contextual relationships and interactions. They believed their findings supported "the existence of survival values associated with flock organization that may, if they exist in nature, give a direct basis for natural selection of socially organized flocks."[18] The group, they reasoned, was sufficiently integrated to serve as a unit upon which natural selection could act.

Thus dominance, despite the fact that it arose as a result of individual aggression, could be regarded as having social benefits, when viewed from the context of group advantage. This theme of group advantage, which was itself connected to notions of group physiology and selection, was a theme most systematically developed at Chicago. By focusing on the group level, dominance and subordination could be regarded as mechanisms of social coordination and in this way be construed as having cooperative benefits. Once the dominance-subordination hierarchy was established, for example, there was little intragroup aggression. As Allee's student, Bernard Greenberg, argued, repeated attacks on the subordinate helped release the tension of high-ranking individuals, creating a greater degree of group integration.[19] Hence, competition, by establishing patterns of dominance and subordination, led to a cooperative, socially organized group. As Allee and Guhl argued, "individual-against-individual competition, such as brings about the peck-order type of social organization of flocks of hens, may serve to build a co-operative social unit better fitted to compete or to co-operate with other flocks at the group level than are socially unorganized groups."[20]

In exploring the physiological and psychological basis of behavior, Allee underscored the importance of experimental analysis and technique, an approach most often associated with the laboratory. Dominance hierarchies were not, however, the only type of social organization found in the vertebrates. When Nikolaas Tinbergen published *Social Behavior in Animals* in 1953, he criticized the American literature for overemphasizing the importance of peck order as a principle of social organization.[21] Tinbergen's comments point to the laboratory bias of Chicago behavioral studies, for it was in the laboratory that the peck order received its greatest verification. Only a handful of studies existed in the early 1940s that documented the presence of dominance hierarchies under natural conditions.[22] In contrast, numerous studies undertaken primarily by ornithologists and amateurs had documented territorial behavior in the field.[23] Margaret Morse Nice, a pioneer in the study of behavioral ecology and an avid promoter of continental ethology in the United States, pointed out the similar shortcomings of Chicago studies in her review of Allee's *Social Life of Animals,* published in 1938. Although Nice thought it an "interesting and well-written book," she cautioned that while the "ornithologist will find much of value on fundamental viewpoints and experiments in the laboratory," there was "little on the behavior of wild birds."[24] Allee did direct a number of experiments on the relationship between territoriality and dominance, which suggested that dominance was, in fact, a function of limited physical space.[25] Still, he refused to entertain the possibility that dominance was merely a laboratory artifact as Lorenz suggested in his critique of G. K. Noble's work.

Upset by remarks that he had overemphasized the importance of peck order in vertebrate societies, Allee wrote to Tinbergen stating that although he had for years "been amused by the peck-order studies," these were "much less significant than studies dealing with survival values associated with optimal crowding." "I have steadily subordinated discussion of peck-orders," Allee insisted, "in comparison with the emphasis put on what I have come to call the proto-cooperative aspects of biology; which can be called group survival value, physiological facilitation, unconscious mutualism, or non-conscious cooperation."[26] Allee's explanations of dominance hierarchies bear many similarities to Emerson's discussions of levels of integration and the population as a superorganism. Both emphasized the population as a unit of selection. Both saw competition as a cooperative and disoperative force. And both saw dominance and subordination as important integrating mechanisms leading to social stability. Yet, as his comments to Tinbergen suggest, Allee did not place a high value on his dominance-subordination research. Even though the dominance hierarchy provided some group advantages, Allee ultimately rejected social dominance as a viable integrating mechanism for achieving social stability in human society. The fact that he emphasized the importance of environmental causes in the development of aggressive behavior indicates the extent to which he hoped that aggression and dominance could be effectively minimized in human society. Furthermore, the hierarchical concept of power and the role individual competition played in the structuring of dominance hierarchies conflicted with his concerns for a politics of peace.

Cooperation as the Means and End

Allee's associations with the American peace movement continued unabated during the interwar years and throughout World War II. In the 1930s, peace activists often devoted their efforts to education, particularly with respect to world affairs. The American Friends Service Committee was a leader in educational outreach programs. Under the direction of Ray Newton, the AFSC Peace Section organized a series of summer institutes on international relations that instructed participants in the economic, political, scientific, social, and religious aspects of peace.[27] Northwestern University was an important Midwest center for these institutes, hosting summer conferences in 1932, 1934, 1935, 1936, and 1937. Allee attended the Midwest Institute of International Relations on at least two occasions. In 1932, he gave a course on the biological background of peace, and in 1935, he gave a special lecture, "Some Biological Aspects of International Relations." He was also a member of the Social-Industrial Section of the American Friends Service Committee (AFSC) from 1936 to 1948, a section dedicated to organizing relief programs and promoting worker education, especially in the coal-mining regions of the Allegheny district.[28] As part of this program, the AFSC sponsored summer work camps, in which college students would help build homesteading communities for displaced miners. In her novel, *The Camp at Westlands,* Marjorie Hill Allee captured the life and motivations behind these community work experiences. The work camps provided young adults valuable training in manual skills and experience in cooperative living, and, at the same time, helped the homestead projects. Westland was a shared farmstead, occupied by fifty families whose primary livelihood came from mining. Through cooperative ventures, the community hoped to protect itself from the domination of the mining industry by diversifying its economic base. As Don Warren, the homestead manager in Hill Allee's novel explained when asked about the unions: "Here, where a man had his garden and his chickens to fall back on, he was not driven to make so quick a bargain for his labor. That might help. So far the struggles between unions and mine owners had not touched Westlands seriously."[29]

The AFSC's emphasis on worker education was connected to widespread concerns with labor reform shared by many pacifists during the interwar period. Violence and authoritarianism were denounced by pacifists as legitimate instruments of social control. Labor was an important avenue for social change because workers represented an economically and politically oppressed group within American society. Norman Thomas, a leader in the Socialist party and active member of the Fellowship of Reconciliation (FOR), a group dedicated to Christian pacifism and the use of nonviolence in social action with whom Allee had affiliations, criticized capitalism because it created an inequitable distribution of power by limiting labor's strength. Allee expressed similar concerns in a commencement speech he delivered to the graduating class of his alma mater, Earlham College, in 1934. Fearful of authoritarianism, Allee saw a "national drive toward regimentation" in Roosevelt's New Deal policies. America was heading toward a dictatorship, and "the underlying forces tending to forge an American fascism, a dictatorship of conservatives" Allee found "lying close at hand." "American business is organized," he stated,

> on the principle of dictatorship. In big business there is no longer a free interchange of ideas between employer and employed, such as marked an earlier and simpler American scene. There is rather the overlordship of the business executive, with a line organization built to carry without delay ideas and orders from the chief down to the humbler workers, but which conveys suggestions in the opposite direction only with much friction. It is too much to expect a minor employee of a large corporation whose every pay check depends on his more or less intelligent subservience, but at all cost his subservience, to a company authority; I repeat, it is too much to expect a man with such a training to show himself outside of working hours a courageous and resourceful supporter of American rights and liberties.[30]

Business in corporate America was structured according to a dominance-subordination hierarchy, a model that denied power to the working class. The business executive dominated and controlled the workers just as the *alpha* individual in domestic fowl subjugated the entire flock. Allee embraced a vision of social democracy, a democracy freed of authoritarianism and violence, a democracy created by voluntarism and cooperation. His comments on the "overlordship" of the business executive in his Earlham address is indicative of his commitment to a strict egalitarian view of economic and political power. In this regard, his remarks paralleled Thomas's evaluation of American capitalism during the Great Depression.

During the late 1930s, Allee became more and more preoccupied with problems of international relations as the threat of war increased. In 1938, he published *The Social Life of Animals,* a popular account of his studies on animal aggregations and researches on peck order that grew out of the Norman Wait Harris lectures he delivered at Northwestern University in 1937. In a chapter entitled "Some Human Implications," Allee maintained that the international political system was structured around war. The focus of the chapter was twofold: to examine whether war was biologically justified and to see whether a system of international relations could be built on the principles of both struggle and cooperation. Much of the chapter was taken verbatim from Allee's 1935 lecture to the Midwest Institute of International Relations.

Allee began the chapter with the question whether humans were biologically endowed with an "instinctive drive toward war."[31] Citing a 1935 poll taken among American psychologists, he tentatively suggested that current scientific opinion supported the claim that war was a learned behavior pattern. Still, other biological factors might be operative in the generation of war. Recalling discussions immediately after the First World War that implicated overpopulation as a cause, Allee reviewed the literature on population growth. He argued that human growth rates were slowing down and that developments in agricultural science were sufficient to increase the food supply to feed a world population ten times its present size. Even if these arguments were wrong, however, Allee went on to show that war was not a biologically sound mechanism for reducing population size by rehashing much of the antiwar eugenics literature written by biologists like David Starr Jordan and Vernon L. Kellogg during World War I. Given that war was biologically unsound, what model of international relations should prevail? Allee argued for an international system of "representative government, a relatively unbiased court of final judicial appeal, and certain police power."[32] He eschewed the League of Nations because it was essentially a league of victor nations and dominant powers intent on preserving their own interests. He favored instead the World Court as a model for an international judicial system. Reluctantly admitting the need for an international police force that would enforce judicial decisions when necessary, Allee forewarned that "if an international organisation is to succeed, police power must be used very rarely. . . . The attempts of the British government to coerce the American colonies or the Irish people," he reminded readers, "are conspicuous as a demonstration of the frequent failure of massed force to compose complex human maladjustments."[33]

Allee hoped for an international system built on "the method of international co-operation based entirely on patience, wisdom and justice."[34] We needed to dispense with our narrow, petty notions of cooperation and struggle, he insisted, and place them "in their proper relation on a world-wide and species-wide basis. . . . The essential struggle of man," was not, according to Allee, "the class struggle of worker against overlord, . . . the race struggle of color against white, . . . the international struggle on however grand a scale . . . the essential human struggle remains the struggle of mankind against his environment."[35] By directing attention to environmental struggle, to the problems of famine and disease, Allee felt that humanity would come to realize the powerful role cooperation played in bringing about social change.

With the U.S. entry into World War II, Allee focused more diligently on ensuring that the biological message of peace and cooperative international relations be heard. Through teaching and his membership in the 57th Street Meeting of Friends, Allee came in contact with numerous young adults who sought advice on draft registration. "Soldier boys and conscientious objectors in about equal numbers seem to haunt our house nowadays," Marjorie wrote her daughter, "with a good number of zoologists still."[36] In the spring of 1940, Margaret Morse Nice brought a young biologist by the name of Joost ter Pelkwyk to Allee's lab. Pelkwyk, a student of Tinbergen's at the University of Leyden, was passing through Chicago en route to Dutch Guiana when the Netherlands was invaded. He decided to remain in the United States and work toward a doctorate with Allee on the social behavior of fishes.[37] Allee expressed a great fondness for Pelkwyk. They spoke not only of shared biological interests, but Pelkwyk also found in Allee someone to confide in regarding his "own strong love for peace." In the fall of 1940, Pelkwyk decided to relinquish his Ridgeway Fellowship, and registered for noncombatant service with the Netherlands government, working as a fisheries biologist in Java. Allee was deeply saddened when he heard in 1945 that Joost had been killed by Japanese gunners in March 1942. He saw in Joost a young man, not so different from himself when he served as a conscientious objector during the First World War.[38]

Allee also helped establish organizational links between biologists and others interested in peace issues and the social dimensions of their science. C. V. Taylor, a protozoologist at Stanford and a close friend of Allee, for example, wrote for advice on organizing a symposium on the biological basis of war and peace for the Stanford semicentennial celebration in 1941. Allee endorsed Taylor's plans, although he did feel that "such a conference should begin largely at the level of the ecological community or population rather than at the level of the physiology or development of individuals," as Taylor had suggested.[39] Immediately after the war, as talks of world government and international cooperation were voiced before the chill of the cold war set in, Allee's reputation as a spokesperson within the biological community for the principles of cooperation continued. The American Society of Zoologists and the zoology section of the AAAS tried to convince him to organize a symposium on the biological basis of cooperation for the centennial meetings of the AAAS in 1948. Allee declined, but Thurlow

Nelson, vice-president of section F of the AAAS, urged him at least to participate in the panel discussion. "Our fellow zoologists would take such a message from you," Nelson replied, "first because of the research which has been going on in your laboratory for so long a period, and, second because of what you are yourself. . . . Too long," Nelson continued, "has the world listened to scientists like the present head of UNESCO, Huxley, and been told that there is no message from biology that will help men to live together in peace." Despite Nelson's passionate pleas, Allee could not be swayed. He did, however, recommend a number of people, mostly from Chicago, who could preach the cooperative message of the biologist. These included Ralph Gerard and Willis Johnson; Emerson was at the time out of the country.[40]

Apart from the general advocating of the value in cooperation, more concrete discussions, especially among Stanford and Chicago biologists, centered on formulations for a postwar international peace plan. Joseph Needham, a biochemist and active member of the British Association of Scientific Workers, met with the biology staff and others at Stanford in May 1942 to discuss his proposal for a world peace organization. C. V. Taylor forwarded Needham's proposal to Allee, who then sent it to E. Raymond Wilson, a former AFSC leader and member of FOR, for possible suggestions but received little response. Allee's own reaction to Needham's proposal was that if such an organization of world peace were to succeed, it needed to put in place a representative legislative body and a judicial world court, as well as a police force subservient to both but used only "after all other solutions have failed." He felt that the "general mind set of mankind" was "toward peace" but that the majority opinion did not often prevail. The task was to "find some way whereby the majority opinion" could "operate," where the drive toward peace could become effective. He likened the situation to his experiments on peck order, where "a strong-driving nervous individual keeps all the rest of the flock stirred up at all times," and where the only effective solution to the problem was to medicate the despotic hen.[41] Allee's call for world government had special appeal for the Chicago-based *Common Cause, a Journal of One World*. The journal's focus was on specific problems of world constitution and government, and Allee was an obvious scientific resource to legitimate their political views.[42] In fact, this was one of the few areas where Allee and University of Chicago president Robert M. Hutchins were in agreement. In July 1945, Allee wrote to Hutchins to express his admiration for a convocation address that Hutchins had delivered in which he expressed his belief in the need for world government. "I entirely approve of your statements and their implications," said Allee. But he was not about to let his disapproval of Hutchin's antinaturalism, his infringements on academic freedom, and his reorganization of the college curriculum pass by. "I only wish that your administration of the internal affairs of the Univ. of Chicago," wrote Allee, "was maintained at the same high level so regularly reached by your public addresses on current national and international problems."[43]

During the war years, Allee chose two highly visible publications from which to address his concerns for international peace and world government. In June 1943, *Science* published his address as vice-president and chairman of the zoology section of the AAAS. "Where Angels Fear to Tread: A Contribution from General Sociology to Human Ethics" evoked such a popular response that it appeared in five other journals over a two-year period: *An Editor's Notebook, Current Religious Thought,* and *Nature* in 1943; *Main Currents in Modern Thoughts* in 1944; and *Sociometry* in 1945.[44] Allee began this article with a summary of his research on dominance-subordination hierarchies and on the cooperative aspects of animal aggregations. These two divergent lines of research indicated the existence of two types of social interactions among animals: "the self-centered egoistic drives which lead to personal advancement and self-preservation, and the group-centered, more-or-less altruistic drives that lead to the preservation of the group."[45] Dominance-subordination patterns were clear examples of individual struggle for social status, while animal aggregations exhibited instances of biological cooperation. Which of these two forces—the egoistic or altruistic—was the stronger? "After much consideration," Allee wrote, "it is my mature conclusion, contrary to Herbert Spencer, that the cooperative forces are biologically the more important and vital."[46]

After presenting the biological evidence for struggle and cooperation, Allee considered the human implications. Like many of his colleagues, he prefaced these remarks with a statement about the negative associations conjured up between science and war. "When the war is over," he affirmed, "the scientists who are now so praised and courted on almost all sides will be told in no uncertain terms, as we have in the past, that the war itself was all our doing."[47] Such accusations would not be completely unjustified. Recalling the antiwar biological rhetoric of the First World War, Allee linked Darwinian evolution with the prevailing militarist philosophy. The biologist, Allee asserted, is "responsible for giving interpretations to some aspects of Darwinian theories of evolution that provide a convenient, plausible explanation and justification for all the aggressive, selfish behavior of which man is capable."[48] Allee was not alone in resurrecting the Darwinist origins of

German militarism. Robert Coker, for instance, who served with Allee on the Committee on the Ecology of Animal Populations, similarly wrote that "an influential portion of some nations has now made the supposed rule of 'the survival of the fittest' in threat and combat the basis of national policy."[49] But if biology could be used as a weapon by the Axis powers, it could also be used as an Allied defense. The question was simply, to what biological vision did one subscribe?

Like Kropotkin, Allee chastised Thomas Henry Huxley for portraying all of nature as the "Hobbesian war of each against all." By separating human ethics from the animal kingdom, Huxley denied that human altruism had its roots in nature.[50] Allee thought otherwise. He took comfort knowing that the "philosophy that condones war is not based on all the biological evidence or on recent interpretations in the light of that evidence."[51] His thirty years of work on animal aggregations provided, he believed, a solid, revisionist view of evolution based on the cooperative associations found throughout the animal kingdom.

Allee ended his *Science* essay with a brief outline for international peace, an outline built around the theme of natural cooperation. He called for relief programs to help undernourished countries, regardless of their political affiliations. In the establishment of a world organization, he urged that the same rules govern both victorious and defeated nations. Hence, all nations should disarm themselves and be equally subject to an international police force. Rejecting political authoritarianism and dominance, Allee also recommended the "curtailment of sovereignty" among both powerful and weak national governments. And reflecting his own experience with the AFSC, he proposed educating "all alike for the processes that make for change by the use of peaceful, non-violent techniques." Only through education and nonviolent action could one build a cooperative world.[52]

Like other members of the American peace movement, Allee adhered to the principle that the end was inextricably linked to the means. A cooperative world could only be achieved through nonviolence and cooperation; dominance and social coercion only resulted in greater social injustice. Writing in the *New Republic* in June 1945, Allee emphasized this point when he remarked that

> sooner or later, however, on the international stage as among our groups of mice, or fish, or hens, or other animals, a subordinate always seriously challenges the *alpha* individual or nation. Although the challenger may be beaten back, often many times, eventually *alpha* rank is taken over by a new despot and the cycle starts again. In so far as any new international organization is

> based primarily on a hierarchy of power, as are the peck orders of the chicken pens, the peace that follows its apparent acceptance will be relatively short and troubled. Permanent peace is not to be won following the precedent established by the dominance order of vertebrate animals.[53]

Allee believed instead that humans should follow the precedents toward association and cooperation that were a fundamental part of all organic life. Because dominance was an important organizational force only in the vertebrate societies, its significance, Allee argued, was less than that of cooperation, which could be found throughout the animal kingdom.

Group organization in vertebrate societies was not based completely on dominance-subordination hierarchies or territoriality. In 1937, F. Fraser Darling published evidence that Scottish herds of red deer exhibited a leadership-followership organization in which an old, experienced female often led herds of hinds and their fawns.[54] Less studied than dominance-subordination hierarchies or territoriality, leadership-followership organizations were important because they displayed a much more flexible type of group structure than those based on dominance. During the 1940s, Allee directed a number of studies to try and determine the relationship, if any, between dominance and leadership in flock organizations. Observations on a group of white Pekin ducks led him to conclude that the most dominant members showed the least signs of leadership.[55] Hence, effective leadership did not depend on coercion, a view that ran contrary to the political philosophy of totalitarian states such as Germany. Still, group organization based on leadership was not the biological utopia Allee envisioned for human society. "The highest organizations," Allee wrote "even tend to dispense with leadership; the individuals composing the group become entirely group centered rather than individually minded. There is neither individual authority nor obedience, for neither is needed in the face of complete cooperation for the common good. This social ideal . . . is the modern social ideal for human society on a world wide basis."[56] The international model for human society was to be derived, not from the peck-order system, nor from leader-follower relations, both of which arose rather late in the evolutionary stages, but from humankind's drives toward cooperation, which existed "even among the Protozoa at an evolutionary level at which only vague premonitions of a possible peck-order system can be detected."[57]

As further evidence for this line of argument, Allee suggested, in a paper presented to a conference on the structure and physiology of animal societies in Paris in March 1950, that if group stability arose out of simple cooperation, such groups would outcompete those based on dominance-subordination hierarchies. "No tests," Allee wrote, "have been made as yet concerning the relative survival value of stable group hierarchies in comparison with wholly pacific groups in which tensions based on differential ascendancy are absent. Although extrapolations are hazardous, particularly in the complex social field, general considerations suggest that the pacific group should make the better showing."[58] Allee's conclusions regarding the superior survival value of peaceful group organizations was purely speculative, but the basis of the experimental design he envisioned had been partially worked out by a former graduate student in the department, John Paul Scott, who was situated at the Roscoe B. Jackson Memorial Laboratories in Bar Harbor, Maine.

The Threshold of Aggression

During the late 1940s, Elizabeth Beeman and Benson Ginsburg, who worked with Allee on aggression in mice, continued their studies on aggression during the summer months at the Jackson Laboratories. The Jackson Labs were originally founded in 1929 by the geneticist C. C. Little to provide pure genetic strains of laboratory animals for biological researchers and to conduct research on the genetic factors affecting abnormal growth. In 1945, however, a Division of Behavior Studies at Jackson Labs was established as part of a ten-year project funded by the Rockefeller Foundation to study the relationship between heredity and social behavior in dogs.[59] The project was first conceived by Alan Gregg, director of the Medical Sciences Division of the Rockefeller Foundation, as the result of a "conviction that one of the constant afflictions of educators is their ignorance of the hereditary equipment of their pupils." Gregg, a close friend and Harvard classmate of the Jackson Lab's director Little, complained that "educators think that environment is everything, but it is not. Consequently a great deal of their effort is wasted or worse. . . . Why would not it be a worth while demonstration, keeping perhaps an occasional animal as a control, to spend 15 or 20 years breeding out an extremely intelligent, relatively small dog just to show that genetically intelligence is capturable and reproducible?"[60] One year later, the Rockefeller Foundation provided $50,000 to Little to launch a program at Jackson Labs and by 1955 had contributed over $500,000 to the project.

Little and Gregg chose as the director for the Division of Behavior Studies John Paul Scott, who was at the time teaching at Wabash College in Crawfordsville, Indiana. Scott received his Ph.D. from the University of Chicago in 1935 for his work in developmental genetics.[61] His adviser, Sewall Wright, had demonstrated that the polydactylous monster that occasionally appeared in guinea pigs was the result of a lethal factor found in a homozygous condition. Scott followed up on Wright's suggestion, tracing the origins of this lethal trait to a critical period in the embryo's development.[62] The notions of thresholds and critical periods in developmental biology later became infused into Scott's behavioral research.

Had Little and Gregg paid closer attention to Scott's earlier writings, they might have chosen a different director for the dog behavior project. Scott's views on the relationship between genetics and behavior were not in keeping with those of Gregg. In directing the program on heredity and social behavior in the dog, Scott never lost sight of the strong tradition in developmental biology at Chicago in which he was indoctrinated. Organisms did not develop according to a simple predetermined plan ingrained in the genetic complex; the process was instead more fluid and represented a complex interaction between organism and environment that was constrained but not determined by the underlying genes.[63] As the project continued, Scott became more and more convinced of the importance of early experience during "critical periods" of the dog's social development. Just as there were physiological thresholds that guided embryonic development from the egg to the wholly formed individual, so were there behavioral thresholds initiated by the environment in the development from the newborn to the adult. The eventual findings did little to confirm Gregg's ideas about the genetic basis of intelligence, and the whole project was buried by the foundation.

Apart from the dog behavior research, however, Scott conducted numerous studies on aggression and social organization at Wabash College and later at Jackson Labs—studies that were funded by the National Institute of Mental Health after 1948. Scott was the person who drew Allee's attention to the difference in aggressive behavior of the three strains of mice bred at Jackson Labs. Building on the study of Ginsburg and Allee, Scott used these mice to determine to what extent social behavior governs social organization. If this were the case, then fighting mice placed together should establish some type of dominance-subordination hierarchy, whereas peaceful mice should exhibit no such organization. Using a single strain, Scott trained the mice not to fight, and then observed the social organization that developed when pacific males were placed together. The mice lived peacefully with one another and no dominance pattern developed. When the same mice were then trained to fight and placed together, an organization based on social dominance emerged. Allee, in his speculations comparing the survival value of dominance hierarchies versus pacific groups, had Scott's experiments in mind. Although he never carried out the experiment, he wanted to take the two different social organizations created by Scott and compare their effects on individual survival and behavior.[64]

Scott extended his analysis of the relationship between social behavior and social organization to a small flock of domestic sheep. Following the general outline proposed by C. R. Carpenter in his influential field study on howler monkeys, Scott classified the social behavior of sheep into eight general types; these included investigation, eating, shelter seeking, fighting, allelomimetic behavior (mutual imitation), epimeletic behavior (care of offspring), et-epimeletic behavior (signaling for parental care), and sexual behavior. Based on the frequency of each of these behaviors and their bearing on the formation of social groups, Scott concluded that allelomimetic behavior was the most significant in determining social organization in sheep. It denoted the tendency for two or more animals to do the same thing as a result of mutual imitation. The leader-follower relations evident in a flock of ewes had their basis in this allelomimetic behavior. Once again, these conclusions suggested that social behavior determines social organization.[65]

Allelomimetic behavior in sheep, like aggressive behavior in mice, was, according to Scott, the result of conditioning and was not based on instinct, as other researchers suggested. The lamb, hearing the calls of its mother, would run to her and be rewarded with milk. From this, Scott hypothesized that the lamb "generalizes from its experience with the mother and thus learns to follow any other sheep."[66] This would explain the general age-structure pattern of leader-follower relations, whereby the oldest ewes tended to lead the flock. To confirm this speculation, Scott isolated two lambs from the flock during the first period of development and observed their subsequent behaviors. Both lambs displayed all types of social behavior except allelomimetic; when released into the flock, they exhibited few tendencies toward social integration.

Scott was quick to point out the significance of these studies for human relations. At the end of his 1945 article on social behavior in sheep, he remarked that there is a

> fundamental belief . . . that the ability to lead must be demonstrated by some sort of competition or fighting. As the small child puts it, "The best fighter is the leader." . . . The observations de-

> scribed above indicate that several different sorts of leadership can exist in the sheep besides that based on fighting, which appeared to have only minor significance under the conditions described. Leadership of the flock went to an elderly ewe, inferior in strength and fighting ability to almost any ram, and often inferior to the younger ewes. This position was achieved mainly by the care and feeding of her descendants without, as far as the observer can tell, any instance of violence toward her offspring.[67]

Following up on this study, Scott went on to analyze the relationship between dominance and leadership in a herd of goats. Like Allee, he found no evidence to indicate the existence of a correlation between leadership and fighting ability.[68]

At the end of his paper on dominance and leadership, Scott cited the studies of the psychologist Harold H. Anderson at the Iowa Child Welfare Research Station on dominative and integrative behavior in preschool children. Interestingly, during this same period, a considerable amount of research on aggression, dominance, and leadership behavior in children's groups was being conducted by social psychologists quite independent of the studies by zoologists in animal behavior. And as in animal behavior studies, the interest was in how differences in authoritarian behavior led to differences in group structure and stability. In each case, there was an attempt to demonstrate that an efficient group structure did not depend on the coercion of individual members of the group by a dominant individual, but that leadership which relied on techniques of cooperation and participatory behavior could result in a stable social organization.[69] One sees in all of these studies an attempt to struggle with the issue of what constituted the ideal political structure for human society and an attempt to refute the totalitarian model on the basis of scientific evidence.

Although Scott was not a pacifist, believing that "there were some situations in which war was preferable to nonresistance," he spoke of his inability to take part in the destructive behavior of war.[70] His age and parental status excluded him from the draft, and during the war years he became preoccupied with the social relevance of biology for human affairs. He always prefaced his articles with such statements as: "The lives of Americans have been dominated for the past fifteen years by war and economic disaster. . . . We need to understand and intelligently control social organization."[71] And he always concluded his essays with a section on human applications. Furthermore, it was during this period that Scott wrote an unpublished manuscript entitled "The New American Destiny" about animal aggression and its relationship to human conduct. This work served as the basis for his 1958 book, *Aggression*, a work that became a focal point of attack for such individuals as Konrad Lorenz and Robert Ardrey, who argued that aggression was an innate survival mechanism among biological species, including humans, and could not be so easily curtailed.

As evidenced by his citation of child psychology experiments in his paper on dominance and leadership, Scott was much more cognizant of the human social psychology literature than Allee. In 1938, he took a leave of absence from Wabash and moved to Boston, reading in the social sciences and attending the lectures of Anna Freud. His position at the Jackson Memorial Laboratories also placed him in contact with leading psychologists. In 1946, for instance, he organized a conference on genetics and social behavior that formally brought together investigators from the fields of zoology and psychology for the first time. Representatives from psychology included Frank Beach, David Levy, Neal Miller, Hobart Mowrer, and Gardner and Lois Murphy. These associations with psychology had a marked influence on Scott's research on aggression.[72]

Scott incorporated into his work one of the most heralded theories in American psychology during the late 1930s—the frustration-aggression hypothesis. This hypothesis, advanced by John Dollard and others at the Yale Institute of Human Relations, was built on the supposition that "the occurrence of aggressive behavior always presupposes the existence of frustration and, contrariwise, that the existence of frustration always leads to some form of aggression."[73] Using the goat flock from his 1945 study on dominance and leadership, Scott created a frustrating situation in which only one animal of a pair could feed at a time. He then observed to what degree dominance and aggression resulted and found a great deal of variation among these pairwise interactions. Once the behavior of these pair relations was established, he increased the time between feedings, thus raising the amount of frustration, and watched for changes in behavior. His experiment indicated that animals that were previously dominant displayed an increase in the amount of aggression when frustration increased, while the subordinate animals exhibited a greater tolerance for repeated attacks. He concluded from this that "frustration tends to cause or at least increase aggression in situations in which animals are in the habit of being aggressive."[74] This did not decrease the significance of frustration as a cause of aggression, but it did, as Scott remarked, "increase the importance of training as a means of controlling aggression, a point which had perhaps best be amplified in connection with human affairs."[75]

While heredity was important in Scott's theory of aggression, it entered "into the picture only in such ways as lowering or raising the threshold of stimulation or modifying the physical equipment for fighting."[76] Individuals thus had different genetic tolerances or thresholds to the amount of environmental stimulation that would invoke aggressive behavior. But aggression was always the result of external conditions. There is, Scott wrote, "no physiological evidence of any spontaneous stimulation for fighting arising within the body. This means that there is no need for fighting, either aggressive or defensive, apart from what happens in the external environment."[77]

In his work on vertebrate social organization, Scott struggled with many of the issues that also informed Allee's research. In their hopes to eliminate future wars, both sought to emphasize the significance of learning and environment in the origins of aggressive behavior. Furthermore, both tried to demonstrate that groups structured around dominance were not necessarily the most efficient and stable social organizations. Yet Scott's early training in physiological genetics, his affiliations with Jackson Memorial Labs, and his interest in psychology led him to emphasize different aspects of animal behavior research than those that characterized Allee's program. Keenly aware of the factors governing individual development, Scott looked to critical periods in the animal's socialization that would influence its subsequent behavior. He was more intrigued with how individual behavior led to established organizational patterns and less interested in the effect that those patterns themselves had on the individual. While Scott focused more intently on the developmental stages of the social organism, Allee chose to focus on the physiological interactions at work in social organizations once they were fully developed. Furthermore, despite their shared interest in the environmental causes of aggression, Scott and Allee had very different interpretations regarding the function of sexual dominance within animal societies. Allee's attempts to develop a theory of sociality independent of sex and the family set him apart from other animal behavior researchers during this period, whose interpretations of sexual and social dominance mirrored their own privileged role within capitalist, patriarchal society.

Integration without Sex

Feminist scholars have repeatedly called our attention to the significance of the dominance and control of "women's sexuality and reproductive capacity" by men in the development of patriarchal societies.[78] The biological sciences have been pivotal in reinforcing gender roles by naturalizing women's assigned social responsibilities, tracing them to adaptations that contributed to the survival value of the species.[79] Donna Haraway, working within the history of primatology, has written extensively on the "logic of domination" that is suffused into, or, rather, identical with the interpretations of primate societies that have appeared in the twentieth century.[80] From Robert Yerkes' work on the psychobiology of chimpanzees to Clarence R. Carpenter's influential study on rhesus monkeys, sexual dominance was considered the integrating force that enabled complex social organizational patterns to appear in the evolution of vertebrate societies.

Within the framework of organic functionalism that marked the terrain of biological and social science discourse in the 1930s, sex and reproduction were the primary organizational gradients used to explain the integration of individuals into the functional wholes known as animal societies. C. R. Carpenter's study of howling monkeys on Barro Colorado Island in the Canal Zone, conducted under the sponsorship of Yerkes and a National Research Council fellowship in the early 1930s, is indicative of the extent to which explanations in animal sociology of the period centered on sex. To understand the process by which group integration took place among howling monkeys, Carpenter broke down group members into three individual categories: adult females (*f*), adult males (*m*), and young animals (*y*). Six pairwise interactions were possible between these three types: *f-f; f-m; f-y; y-y; m-y; m-m*. Of the six possible social relationships, Carpenter considered male-female interactions to be the most important in establishing important social bonds. A "sexually unsatiated" male and a "female in oestrus" were both "forceful incentives" for interrelations to develop. "With the repetition of the reproductive cycle in the female and with uninterrupted breeding throughout the year," Carpenter argued, "the process of group integration through sexual behavior is repeatedly operative, establishing and reinforcing inter-sexual social bonds." Howling monkey societies were distinctive in their "lack of a sharp gradient of dominance and submission"; a female would copulate without any apparent recognition of hierarchical social status among the males. This lack of dominance accounted for the howling monkeys' communal structure. Still, sex was the primary organizational gradient around which Carpenter's explanations of group integration were based. Discussion of the interactions between howling monkeys and their environmental surroundings was limited to a few pages of text.[81]

Although howling monkeys exhibited a low gradient of sexual dominance, the establishment of a colony of rhesus monkeys on the island of Cayo Santiago by Carpenter in December 1938 enabled him to analyze experimentally the significance of sexual dominance as an integrating force within primate societies. Carpenter allowed sufficient time for stable social organizations among rhesus monkeys to be established and then determined the dominance relations among males within a single group named Diablo. After the hierarchy had been determined, Carpenter removed the dominant male from the group and watched the subsequent effects on social organization. The experiment was similar to that performed by Allee and Guhl on social organization among hens in the early 1940s. One week later he removed the next dominant male within the hierarchy; in the third week, he removed the third dominant male. He then returned the males to the group and allowed social relations to become reestablished. The absence of dominant males resulted in a period of social chaos marked by a greater degree of intragroup aggression and by a more fluid organizational structure. In essence, the social organism had gone back to a previous stage of development, before dominance relations had been established that enabled the society to function in an efficient and stable manner. "Carpenter," as Haraway notes, "conceived social space to be like the organic space of a developing organism, and so he looked for gradients that organized the social field through time."[82] In Carpenter's understanding, the main gradient was the sexual hierarchy. Just as Child saw behavioral interactions between protoplasm and environment creating gradient patterns around which organizational structure developed in the growing embryo, so Carpenter viewed sexual relations and dominance among males as the primary organizational gradient that structured rhesus monkey societies.

Carpenter's appeal to dominance as a primary integrating force within animal societies was characteristic of animal behavior studies throughout the 1930s and 1940s. During the Second World War, however, an additional level of complexity was added to the inherently capitalist patriarchal interpretations of dominance, a level of complexity that was accentuated by the threat of totalitarianism abroad. What is striking is that individuals could challenge the notion of a dominative political authority while ignoring the traditional hierarchy of power with respect to gender relations embedded in Western democracy. That individuals could question the merits of domination as a valuable means of integrating human society while still adhering to the legitimacy of dominance within the sphere of sexuality is apparent in the subtle distinctions made between social and sexual dominance during this period. Hence, Scott, in his 1958 book *Aggression,* could point to the biological roots of sexual dominance in humans while denying the validity of social dominance as an important model for leadership during the war years.[83] A similar approach was advocated by G. K. Noble, a leader in the field of animal behavior studies during this period and director of the Department of Experimental Biology at the American Museum of Natural History.

Noble, a son of the noted publisher, Gilbert Clifford Noble, had developed a fascination for studying the life histories of reptiles and birds while an undergraduate at Harvard.[84] He received his A.M. under Thomas Barbour in 1918 and went on to obtain his Ph.D. from Columbia University in 1922. His dissertation, "The Phylogeny of the Salientia," reflected his lifelong interest in herpetology and systematics. Through William K. Gregory, Noble's graduate adviser, professor of vertebrate paleontology and curator of the Department of Comparative Anatomy at the American Museum of Natural History, Noble also developed an appreciation of functional morphology and microscopical anatomy in analyzing phylogenetic relationships.[85]

The close ties between Columbia University and the museum led to Noble's appointment as curator of the Department of Herpetology at the museum in 1924. But Noble was not content with a career devoted entirely to systematics and natural history when the cutting edge of biology lay in the experimental disciplines of neurology, physiology, and endocrinology. A series of offers from Columbia University and Cornell University Medical School in 1928 provided Noble with a bargaining position to meet his own aspirations. In order to stay on as curator at the museum, Noble demanded half of the fifth and sixth floors of the museum's African wing then under construction for a laboratory of experimental biology with an annual budget increase of $10,000 above the department's yearly $17,000 budget. In addition, the Department of Herpetology was to become the Department of Herpetology and Experimental Biology, with half of Noble's time given to experimental biology research.[86]

The museum acquiesced to Noble's demand and in May 1934, the city of New York completed Noble's Laboratory of Experimental Biology at an expense of $78,920.[87] It occupied the entire sixth floor and roof of the African wing and included, among other things, an aquarium room, three greenhouses, an animal house, histology laboratory, and physiology laboratory.[88] But because of the financial constraints caused by the Depression, the museum could no longer maintain its previous level of support. The department's budget for 1934 was cut by $7,000 and slashed another $7,000 in 1935.[89] To adjust for the loss of his research and clerical staff, Noble managed to secure help from the Works Progress Administration. In 1934, he started with seventeen WPA people working in his laboratory; by 1937 this number had escalated to sixty-five. These workers were involved not only in the preparation of exhibits and the maintenance of the aquarium rooms and laboratories. A number of individuals also worked as research assistants. In addition, Noble had a staff of at least seven people responsible for translating biological articles from foreign journals, analyzing the literature dealing with the morphology, physiology, and habits of reptiles, and collecting literature on the courtship and sexual behavior of animals.[90]

Facing budgetary reductions, Noble decided to limit his research to the "physiology and psychology of reproduction in the lower vertebrates" even though the laboratory was originally intended to "consider many problems on the borderline between natural history and biology."[91] He regarded his research on reproductive physiology and behavior as the most significant of his laboratory studies. It was also an area that was being actively supported by the NRC-CRPS, a fact that surely swayed Noble's interests. Having spent the summer quarter at the University of Chicago in 1931, with the hope of borrowing "many ideas and techniques" for his own laboratory, Noble left Chicago impressed by Lillie's large-scale enterprise in reproductive physiology.[92] To be sure, Noble's earlier work on the biology of the amphibia demonstrated an interest in the physiological and psychological problems associated with reproductive behavior. But the NRC-CRPS certainly helped reinforce Noble's interest. An adept entrepeneur, Noble was careful never to distance himself from the public eye, and sex was surely a topic of public interest. As Douglas Burden, a museum trustee, film producer, and close friend, remarked to Noble: "What excitement you will evoke if the substance of your new hormone not only brings immature animals to sexual activity but increases sexual activity among the aged. . . . Pull that trick and we will never have any difficulty raising funds for your research."[93] Taking Burden's comments to heart, Noble went public, publishing an article in the *New York World-Telegram* on "Recent advances in our knowledge of sex."[94]

From 1935 until 1940, Noble developed a program of animal behavior study that utilized the techniques of endocrinology and neural surgery to establish a detailed picture of the mechanisms responsible for social behavior in the evolution of the vertebrates. Unlike Allee, Noble viewed behavior within the context of phylogenetic relationships and instincts, tracing similarities in behavior to common ancestry rather than to environmental relations. His perspective was that of a systematist rather than an ecologist. By analyzing the social behavior of fishes, reptiles, birds, and finally mammals, Noble hoped to ascertain how far phylogenetic changes in neural structure had led to differences in social behavior patterns. For Noble, behavior was to be understood in the neurophysiological structures and processes ingrained in the individual organism as a consequence of its phylogenetic past.

Noble's understanding of social behavior was almost completely oriented around sex. In *The Biology of the Amphibia*, published in 1931, Noble devoted considerable discussion in a chapter on instincts and intelligence to the evolution of courtship behavior in salamanders, detailing how courtship patterns of the various families of salamanders were all modifications of the pattern found in the most primitive group. Regardless of the organism involved, be it salamanders, jewel fish, or black-crowned night herons, the methodological approach of his behavioral studies in the late thirties was identical. He first gathered extensive information on the sexual behavior patterns of the organism in the wild, next performed an analysis of hormonal effects on these behaviors in the laboratory, and then used ablation techniques to isolate the neural centers involved. In this way, he hoped to compile a phylogenetic history of the evolution of courtship behavior from fish to humans.[95]

When Noble died suddenly in 1940, at the age of forty-seven, from a streptococcus infection known as Ludwig's quinsey, the majority of his research assistants applied to work with Allee. Lacking the institutional support Noble had, Allee turned all but one student, Bernard Greenberg, away. Allee's laboratory was the most obvious institution in the early 1940s to which young animal behavior researchers trained under Noble might apply. Yet, the research programs under these two investigators, despite their shared interest in social behavior and dominance relationships, displayed distinct underlying differences. One was ecological, the other evolutionary. More important, however, was the role sex played in each of their theories. Noble saw sex as the primary integrating mechanism of vertebrate societies; it was the social bond that kept individuals together and alleviated antagonistic relations. Sex gave birth to the family and society. He took issue with Allee's failure to distinguish between social and sexual dominance. Social dominance, Noble maintained, "tends to keep individuals at a distance. Were it not for the cohesive group attraction, it would completely destroy the group."[96] Thus, for Noble, social dominance was seen as a destructive rather than integrative force. Sexual dominance, on the other hand, made the bonding of mates possible and was thus a positive force in group attraction. Indeed, in Noble's illustration of the courtship behavior of the black-crowned night heron, the consummatory act of copulation can only be secured after the male has gained a dominant head position over its mate.[97] Hence, by distinguishing between social and sexual dominance, Noble and Scott could challenge the authoritative structure of totalitarianism, while still reaping the privileges of male dominance within a democratic society.

Allee's affiliations with the liberal pacifist movement, however, sensitized him to the issue of hierarchical power relations, be they based on gender, race, or class. He did not distinguish between sexual and social dominance and denied the validity of any form of domination as a viable integrating mechanism in human society. In certain respects, his own personal life transcended many of the traditional boundaries that structured family and gender roles in early twentieth-century American society giving clues to his attitudes toward gendered hierarchies. He and his wife, Marjorie, shared a marriage in which both pursued professional careers. Furthermore, Marjorie took an active part intellectually in Clyde's professional life. He often brought his professional correspondence home to her during lunch, where the latest news and ideas in the biological community would be read and discussed together. Marjorie also played an important editorial role in the writing of Clyde's professional articles. In addition, over 40 percent of Allee's thirty-two Ph.D. students at the University of Chicago were women, a high percentage in a department where the average number of doctorates granted to women between 1920 and 1950 was 30 percent.[98]

Allee's theory of sociality was not rooted in sex or the family, but in a political economy of peace that stressed the role of cooperative associations in nature, associations formed to alleviate the severity of struggle between the organism and its environment. Dominance was a late act in the evolutionary drama of life. A more rudimentary integrating social force existed, first expressed in the tolerance individual organisms displayed toward one another when they gathered together as a consequence of similar reactions to environmental conditions. This nonconscious cooperation, Allee argued, was the wellspring from which sociality sprang and to which human society should look as a biological pattern for social life.

The birth of culture through sex and the family, which has been the mainstay of nineteenth- and twentieth-century scientific theories on the origins of society, legitimizes a hierarchy of power by describing the nuclear family as an evolutionary advance, in which a sexual division of labor, integrated through the male's "natural" position of dominance, created a more efficient, stable social unit. The formation of societies is thus attributed to innate biological differences between the sexes, a view which has often been extended to include race, thereby giving natural sanction to the domination of non-Western peoples by the white, Anglo-Saxon male. Yet Allee, in ascribing the emergence of sociality to the interaction between organism and environment, obfuscated the division of labor/hierarchy concept at the heart of biological/social science discourse. In Allee's views, sociality did not arise through integration of innate differences but was a property shared by all living organisms, in which differences were sublimated in the common struggle with adverse environmental conditions.

This theme of union and cooperation against adversity pervaded not only Allee's biological ideas but was a common thread in his reform activities and in the writings of his wife Marjorie Allee. In the early 1940s, Allee became involved in the establishment of interracial summer work camps in the south end of Chicago. Individuals from diverse racial backgrounds lived together for the summer in residences located in zones of racial tension. The aim was to "break down racial discriminations in day-to-day living," but the threats by individuals within local neighborhoods opposed to such racial integration were often severe. Restrictive covenants that prevented blacks and Asians from renting or buying properties were common in communities throughout Chicago, before a Supreme Court decision declared them illegal in 1948.[99] Allee described one instance in the summer of 1945, when a house chosen for the summer work site was ransacked after locals learned of the "owner's decision to rent the house to a mixed group." The incident only affirmed in Allee's mind that "the community chosen for the work camp needs such attention rather badly. The challenge would be real." The challenge was important, for it was through adversity that individuals recognized their common experiences, shared needs, and the mutual benefit to be gained through cooperative life.[100]

The interracial summer work camps, in addition to her own daughter's experiences in an AFSC co-op in Philadelphia, provided Marjorie Allee with the backdrop for her 1944 novel, *The House*, which won her an award from the Child Study Association of America.[101] *The House* takes place in Chicago, where a group of young adults, some of whom are graduate students at the University of Chicago, venture into an experiment in cooperative living. Early in the novel, the group faces the reality of racial discrimination, when a visit by the neighbor Miss Johnson informs them that the older residents of the community have expressed dismay over the frequent visits by an Oriental woman, Alice Chen, and a black woman, Delinea Johnson, to the young people's home. The group had been considering Alice as a possible member but placed her on the waiting list as a result of Miss Johnson's warnings. Shortly afterwards, however, the venturesome band was evicted. Hardship is once again turned into advantage, for the group finds a new home more suited to their needs in an interracial neighborhood where Alice is free to live as a house member. The rest of the novel details the trials and joys of group living and the omnipresent grip of war on everyone's lives. In the final pages, Merritt Lane likens the house to a "community of savages, working under public opinion freely expressed, without laws or law courts," as described by Alfred Russell Wallace in *The Malay Archipelago*. "In such a community all are equal," she continues, "there is none of that widespread division of labor which produces conflicting interests."[102] Here, one finds an expression of the idealized vision of social life that structured W. C. Allee's ideas about the kingdom of nature.

Merritt Lane and Delinea Johnson made their first appearance in an earlier novel written by Marjorie, *The Great Tradition*. The setting in *The Great Tradition* is the University of Chicago's zoology department, and the main characters are all women students pursuing biology degrees. Delinea is a black woman, a dean of women at a small, financially destitute, black college in Pennsylvania, who has enrolled in the department to obtain her master's degree in order to teach at a more prestigious black institution. Delinea's experiences bear unmistakable similarities to those of Roger Arliner Young, a research assistant of the biologist Ernest Everett Just, who received his Ph.D. from Chicago and taught at Howard University. Allee supported the hiring of Just as an instructor in the Chicago zoology department in 1935, but Just was rejected as a possible candidate for racial reasons.[103] Racial prejudice is also a subplot in *The Great Tradition*, played out between Delinea Johnson and another woman graduate student, Charlotte White, who comes from a prosperous family in Mississippi. Marjorie Allee portrays the fractured social worlds which Delinea must face: her dignified status as a woman dean within the black community; her subservient role as maid of the household in which Merritt and Charlotte live; her struggles as a poorly prepared graduate student, a victim of unjust educational and economic conditions. And Charlotte sees Delinea only through the lens of deep-seated racial prejudice. "No nigger," Charlotte declares in one of her many diatribes launched against Delinea, "has a place in a university where research is carried on." But the walls of Charlotte's prejudice begin to crumble when Delinea sacrifices herself—injuring her eyes in the process—to protect Charlotte from a laboratory explosion caused by Charlotte's negligence. Resolution once again comes in the shared strife against harsh external circumstances.[104]

Just as Marjorie Allee crafted her stories to end in harmony and cooperation, so Clyde Allee, as John Beatty has aptly remarked, "constructed a world in which eternal peace and order were possible."[105]

His research on dominance-subordination hierarchies indicated, to him at least, that dominance arises rather late in the evolutionary stages of sociality and could thus be subverted. Furthermore, by invoking group selection as an explanation of the dominance hierarchy, he tried to accommodate the presence of competitive interactions within his cooperative world. In restructuring the economy of nature, Allee was also reshaping the political boundaries of human society. His ideas threatened the accepted construction of biological/social theory that gave credence to twentieth-century American capitalist society; yet, like his colleagues, Emerson and Gerard, he looked to the natural economy as a legitimation of his own political views.

9

Redefining the Economy of Nature

One year after the 1949 publication of *The Principles of Animal Ecology,* Allee, at age 65, retired from the University of Chicago and accepted a position as head of the Department of Biology at the University of Florida in Gainesville. His longtime hope of chairing a department was fulfilled, but his stay at Florida was fraught with controversy as he sought to restructure a large and highly inbred department. Five years later, on 18 March 1955, Allee died from complications associated with a severe kidney infection. In the last decade of his life, Allee witnessed an important transitional period in the cultural history of the life sciences within America, a period that by the 1960s had brought the coup de grace to many of the ideals so central to the Chicago ecology enterprise. By 1949, appeals for a cooperative world order had rapidly faded, displaced by Cold War fears of the Communist threat both within and outside U.S. borders. Even the 1950 statement by UNESCO that "man is born with drives toward co-operation, and unless these drives are satisfied, men and nations alike fall ill," was absent from the 1951 revision.[1] The appeal to cooperation could no longer be sustained within the framework of the modern synthesis, especially as this framework hardened to a rigid adaptationist program where evolution occurred through natural selection of *individuals* within a population.[2] Nor could the role of cooperation be seriously entertained beyond the boundaries of the nuclear family, for such appeals of group solidarity resonated too closely with the ideology of totalitarianism, which had not only threatened the existence of democracy in the Second World War, but by the late 1940s had become increasingly identified with the Soviet state.

The early sustenance of Chicago animal ecology within a developmental, organismic framework was essential for the maintenance of the various social guises within which Chicago ecology was dressed: the liberal capitalism under "some strong social control" espoused by Emerson and the more radical, social democratic antiauthoritarian ideals of Allee.[3] In building an animal ecology informed by experimental embryology and physiology, Allee developed a model of sociality in which cooperation could operate as a fundamental organizing force. The development of the community and, hence, of society was identical with the development of the organism from the egg to the adult. Each cell or individual was part of the body physiologic or the body politic. The individual, in its interactions with the environment, contributed to the organizational pattern of the functional whole, and through its participation in the social organism received benefits unavailable in a solitary life. Differences or conflicts of interest were thus submerged in the complete cooperation for the common good. Emerson helped recast Chicago biology in a Darwinian mold, melding a hereditarian perspective centered on natural selection with a physiological understanding of what constitutes the organismic unit of evolutionary change. Yet as Allee reshaped Chicago animal ecology to include an evolutionary vista, a move most evident in the shift from the community to the population as the primary unit of ecological study, he introduced tensions in the landscape that would seriously jeopardize the legitimation of his social vision. The social organism was retained as Allee studied the population in organicist terms, yet the development of the social organism was no longer solely the consequence of ontogeny. Instead, survival and reproduction of individuals, in short, heredity, played an important role. By introducing elements from a neo-Darwinian perspective, Allee also introduced the possibility for conflicts of interest to emerge between individual members of the group. He and Emerson tried to ensure that organisms acted harmoniously together by appealing to group selection, but this could easily be stripped of its organicist context to reveal a world governed strictly by competition and the search for individual success. The boundaries of the individual, in this instance, the population, could no longer be identified solely in relation to the external environment, as was the case with the community. Instead, the notion of a shared genetic complex would have to enter in. The locus of individuality had shifted in postwar ecology and evolutionary biology, and over the next two decades, that locus would become increasingly identified with the individual gene.

George C. Williams's sharp 1966 rebuttal in *Adaptation and Natural Selection* of V. C. Wynne Edwards's 1962 book, *Animal Dispersion in Relation to Social Behavior,* brought the debate over the appropriate unit of evolutionary discourse to the fore. Williams attacked the notion of group selection by arguing for a strategy of parsimony in which the fundamental unit of selection was not the species or population, but the individual gene.[4] Hence, as the domain of the individual became more and more restricted within evolutionary theory, a move that was decisively connected to the growing hegemony of genetics and molecular biology within the life sciences, the importance of cooperation as a theoretical construct waned. As Evelyn Fox Keller has remarked, "the bigger, the more complex the individual, the more scope there is for internal cooperation, interdependence, for functional and/or purposive dynamics; the smaller the individual, the larger the scope for external competition and/or random interactions between individuals."[5] Since the population is itself interpreted as an amalgam of individuals with distinct genetic identities, cooperation is not a viable organizing force, except in situations in which there is a high proportion of shared genes between individuals. Altruism is actually a selfish game, played only when the individual can expect a return greater than its original investment. The population is no longer understood as an organic whole, in a developmental sense, in which all the individuals are an integral part. Individuality is defined, first and foremost, by hereditary characteristics, and the genetic differences between individuals constitute an essential component of evolutionary change.

What constitutes the boundary of the individual within the biological sciences is a highly contested point. As Emerson noted one year before his retirement from the University of Chicago in 1962:

> Some can see the borders of an individual, but fail to recognize the borders of a species, and some who recognize the species may be unaware of the interspecies community system. Contrariwise, some biologists who are fully conscious of the assemblages of organisms in a coral reef or a tropical rainforest, may treat the individual only as a statistic, and the species as a subjective figment of the imagination of the taxonomist. These differences in perspective are the bases of diverse political philosophies now plaguing human society and international relations. Overemphasis upon the state to the exclusion of the individual freedom, or overemphasis upon the rights of the individual to the exclusion of his social responsibilities are the ideological causes of both cold and hot wars.[6]

Emerson's comments suggest that much is politically at stake in locating the boundaries of the individual within ecological and evolutionary discourse. For him, those boundaries served as a demarcation between the individual and the state. In Keller's words, the individual serves as a demarcation between public and private life. In the public sphere, Keller writes, one has "autonomy, competition, simplicity; a theoretical privileging of chance and random interactions, and the interchangeability (equality) of units." In the private sphere, one has "interdependence, cooperation, complexity; the theoretical privileging of purposive and functional dynamics, and often a hierarchical organization."[7] Yet our understanding of what is the appropriate boundary between public and private life is not static and absolute, but ever-shifting, highly dependent on personal values, political beliefs, and cultural norms.

A consideration of attitudes in postwar America about the relationship between private and public space, between the individual and the state, offers intriguing yet admittedly speculative insights into the demise of a natural economy rooted in organicism and cooperation. George Gaylord Simpson's arguments against the "aggregation ethics" of Emerson and Gerard serve as an important inroad into the landscape of concerns that marked 1950s American culture, a landscape that may well have influenced the direction of ecology and evolutionary biology in the postwar years. Simpson was fearful that organicism could only lead to the subordination of the individual. This fear took on increased proportions as the specter of totalitarianism in the Soviet Union under Stalin loomed large with the onslaught of the Cold War. One of the features that defined the ideological conflict between the United States and the Soviet Union in the late 1940s, as historian Edward Purcell notes, was the belief among American intellectuals that "absolutist philosophies, such as Marxism, inevitably led to totalitarianism, regardless of the other political and social institutions of a society."[8] Subservience to the group under Communist regimes represented a dangerous foreshadowing of absolutist and, hence, totalitarian philosophy. The United States, however, with its pluralist society of diverse and competing interest groups was, by its very institutional nature, sheltered from the threat of dictatorship. Difference, diversity, and conflict were the preserving forces of a democratic society; cooperation and group solidarity were not.

Community was a powerful metaphor for Allee and other liberals during the interwar years, but by the 1950s its meaning had shifted. No longer did it provide the individual with a sense of meaning, purpose, and value. On the contrary, community, with its connotations of conformity, threatened to undermine individual identity. As David Riesman, professor of social science at the University of Chicago, noted in his 1951 essay, "Individualism Reconsidered," the "danger of the 'garrison state'" in postwar America was much greater than anarchy or "unbridled individualism."[9] The threat of cutthroat individualism reminiscent of the Victorian age had long since passed; managerial capitalism, with all its benefits, was here to stay. In the affluent years of the 1950s, the structure of American society went unchallenged. Instead, people began to turn inward, to a "quest for identity and self-fulfillment" in the words of Richard Pells.[10] Riesman, for example, encouraged "people to develop their private selves—to escape from groupism—while realizing that, in many cases, they will use their freedom in unattractive or 'idle' ways." But the very prosperity of American society, Riesman argued, made it possible "to take the minor risks and losses involved in such encouragement as against the absolutely certain risks involved in a total mobilization of intellect and imagination." Encroachment of the state on the individual had become much more important than the sacrifice of individual identity to the public self. "No ideology," Riesman insisted, "however noble, can justify the sacrifice of an individual to the needs of the group."[11]

Participation in civic life and in social causes had become much less of an issue in 1950s America than it had in the interwar period; the search for self in the midst of a mass consumer society at home and totalitarianism abroad became a more central concern. In Allee and Emerson's organicist views, a harmony existed often, though not always, between the ideals of the individual and those of the group. In the postwar years, however, the boundaries of the private sphere had come to envelop the individual person alone. Neither the community, nor the state, for that matter, were deemed a part of private life. Indeed, the resistance of the individual self to the intrusive pressures of the group was considered vital to the preservation of democratic life, so long as the existing institutions within America were not threatened. Jack Finney's 1955 classic science fiction novel, *The Body Snatchers,* illustrates this theme well, for it is the "inner-directed" individualism of the hero Miles Bennell and his resistance to conformity that saves the world from the complete obliteration of individual personality in a mass society taken over by an alien life form.[12]

Diversity and conflicts of interest created by the freedom of the individual to choose to conform or not "to the power-requirements and conventions of society" ensured that power within the American democratic system was dispersed.[13] "Our very gifts . . . in America," Riesman reasoned, are "our ability to move in different directions, to be *unintegrated* to a degree, to operate on discontinuities of life and career."[14] Within the fractured world of competing groups, the individual was free to choose his or her own way. And that choice was to be based on personal advantage and self-interest. Even cooperation, Riesman argued, should be secured from individuals by "offering concrete advantages—including such personal growth as they may gain from the co-operation itself—without asking them what their fundamental values are." One cooperated with others, not because it was in the interest of the group, but because it was in one's own interest.[15]

In the partial shift from a developmental, physiological emphasis to a neo-Darwinian frame, Allee and Emerson preserved the continuity of cooperation by adhering to a group selectionist account in which the population was the organismic unit. Individuals acted for the good of the group. Yet, the tenets of organicism and cooperation embedded in their ecological and social views became much more difficult to sustain amid postwar anxieties about "groupism" and the loss of individuality. Park, in contrast, passed through the postwar period unscathed. By the early 1950s, his own population studies had become divorced from the cooperationist, physiological framework of Allee's research program. Although he titled his early papers studies in population physiology, by the late 1940s such physiological references were notably absent.[16] In fact, his laboratory studies of competition in *Tribolium* have been heralded by other ecologists as contributing to "the demise of the superorganism."[17]

As the domain of private life, centered around individual choice and freedom, became more revered in the 1950s, so too did the value of competition. To act out of self-interest meant necessarily that conflict would arise. Such competition, however, did not lead to chaos and disorder. On the contrary, as Robert Dahl suggested in his 1956 work, *A Preface to Democratic Theory,* competition, by increasing "the size, number, and variety of minorities whose preferences must be taken into account" in the political decision-making process, prevented power from being concentrated in the hands of a few. Competition between a diverse number of minority groups coupled with political consensus were the key features that distinguished democracy from dictatorship and preserved the stability of the democratic system.[18]

Such prevailing cultural concerns with conformity and individualism, coupled with developments in genetics and theoretical population ecology, provide a useful frame for understanding the redefinition of nature's economy in postwar ecology. David Lack's 1954 book, *The Natural Regulation of Animal Numbers,* signaled an early warning that the population, viewed as a social organism, was dead. Population properties such as clutch size, Lack maintained, could be explained solely as a consequence of the evolution of reproductive tactics for achieving individual success.[19] Competition between individuals, rather than cooperation among individuals for the maintenance of the population, became the major force structuring the economy of nature. And just as competition increased diversity and led to stability within the political economy of the United States, so too in the economy of nature. In the mid-1950s, Robert MacArthur advanced the diversity-stability hypothesis within population ecology, in which he suggested that the stability of species populations was greater in more diverse communities. In areas such as the tropical rain forest, with rich species diversity and intense competition, a host of checks and balances operated, MacArthur argued, to keep population sizes of a given species within a narrow and, hence, stable range. Although this hypothesis was later criticized by Robert May and remains controversial within theoretical ecology, it does point to the central importance that competition as a creative and stabilizing force within nature's economy began to play in ecological thought in the immediate postwar period.[20]

Once competition between individuals became the primary explanation of community structure, and individuality was defined by reference to genetic identity and distinctiveness, the influence of group organization, of environmental context, on individual behavior became less significant. Social organization such as territoriality was now merely an outcome of individual behaviors and interactions, nothing more. In the period of economic prosperity that followed the Second World War, a period in which democracy was itself heralded as the ideal form of political organization, the individual organism, as carrier of genotype, was suspect as the harbinger of social problems. Juvenile delinquency was one social ill that did not disappear in affluent American society. But the cause of such antisocial behavior could not be blamed on the structure of social institutions. As animal behaviorist David E. Davis suggested, the origin of juvenile gang behavior was far "more profound than merely a demonstration of adolescent male adjustment to society." Aggressiveness, Davis argued, was innate, "heavily dependent on genetics." Individuals may have more or less genetic propensity to fight; but, Davis urged, "only the means of fighting and the object of attack are learned."[21]

Davis's article on the phylogeny of gangs was but one of many works that appeared in the ethology and animal behavior literature during the early 1960s, emphasizing the innate origins of aggression in human society, which culminated in the published English translation of Konrad Lorenz's *On Aggression* in 1966.[22] In *The Territorial Imperative,* Robert Ardrey offered a defense for such appeals to aggression as instinctive behavior. "How could one know," Ardrey pleaded, "that such an age of affluence and material security would witness a level and degree of juvenile delinquency that did not exist in the depression years; racial conflict and bitterness that we had never known; and a crime rate beyond our most monstrous imaginings?"[23] Ardrey's comments may be read as a convenient popular rhetorical justification for his scientific views, but they do signify the extent to which the individual was held responsible for the problems within society. And to avert these problems, the individual had to turn inward, to acknowledge and recognize these genetic predispositions towards aggressive behavior and to find the appropriate social outlets for their release. "Insight into the causality of our actions," Lorenz reasoned, "may endow our moral responsibility with the power to control them," and redirect the "functions of aggression" along less dangerous lines. Sport was one such cultural activity which Lorenz considered to furnish a "healthy safety valve for that most indispensable and, at the same time, most dangerous form of aggression."[24]

To be sure, the pop ethology literature on aggression in the 1960s generated a critical backlash. Titles such as "War is Not in Our Genes" and "Man Has No 'Killer' Instinct" filled the pages of popular magazines and scientific journals such as the *Nation*, the *New York Times Magazine*, and *Science*, challenging the claim that humans are by nature aggressive. In fact, two of Lorenz's most ardent critics, the anthropologist Ashley Montagu and the biologist John Paul Scott, were strongly influenced by Allee's cooperationist writings during the Second World War.[25] Yet what is most illuminating are the underlying similarities rather than the obvious differences between Allee's and Lorenz's outlooks. Both looked to nature and to biology as a way of healing human society. Indeed, Lorenz spoke of ethology as offering "therapeutic" and "preventive measures" to redirect and heal the dangers that lie latent within the human organism.[26] And although E. O. Wilson in *Sociobiology: The New Synthesis* explicitly distanced himself from such earlier ethologists as Lorenz, Wilson still sees biology and nature as offering the normative guidelines along which to construct human society.[27] Ethology and sociobiology continue the tradition of biological humanism advanced by Allee, Emerson, and others during the interwar period, although with a different underlying economy of nature.

While the thread of biological humanism and a naturalistic ethics continues to be pervasive in the study of animal behavior, the appeal to nature for human solace and meaning largely disappeared from the domain of community ecology in the postwar years. Whereas Allee and Emerson looked to the animal kingdom in the hopes of understanding the nature of human society and to biology as a therapeutic for curing social ills, this image of ecologist as healer became more problematic after the Second World War. Within the domain of community ecology, the biosociological approach so characteristic of the interwar period, began to collapse and was replaced by a cybernetic analysis that integrated the biotic and abiotc, nature and machine, into one. While competition became the guiding principle in postwar population ecology, the principles of energy flow and thermodynamics became the basis for ecosystem ecology, which usurped community studies at the forefront of ecological research. Eugene Odum's *Fundamentals of Ecology,* published in 1953, was oriented around the ecosystem concept, and rapidly displaced *Principles of Animal Ecology* as the leading ecology textbook in the immediate Cold War period. Throughout the 1950s, Eugene Odum's Savannah River project on secondary succession and the ecosystem cycling of radioactive isotopes was heavily funded by the Atomic Energy Commission. Ecology not only appropriated military funds, it also appropriated the cybernetics narrative that emerged from military research on aircraft–missile guidance systems.[28] The ecosystem blurred the distinction between inorganic and organic by reducing everything to energy as the common denominator. Nature had become a system of components that could be managed, manipulated, and controlled. The ecologist's task increasingly became that of environmental engineer; ecologists were to be professional managers who could monitor and fix the environmental problems created by human society. This image became especially appealing to bureaucrats and politicians in Washington during the 1960s as they faced a public constituency worried about the accumulation of wastes such as pesticides, phosphates, and radioactive isotopes generated by an affluent, industrialized society.[29]

If, as Daniel Botkin has suggested, we have, as we enter the twenty-first century, "the power to mold nature into what we want it to be or to destroy it completely," what function can "the natural" serve?[30] While the natural once stood in contrast to the artificial, nature has increasingly become artifice, a skillful creation. As developments within the fields of genetics and molecular biology enable some biologists to create new organisms at will, the creation of nature has increasingly become the craft of cell and molecular biologists as engineers.[31] This engineering image has also pervaded the science of ecology in more recent years and is at the heart of the ecosystem concept. Yet once the biologist can invent a blue pig that flies or create and manage artificial ecosystems, how can nature be normative, how can it give definition and purpose to humankind, if nature itself increasingly becomes a human construct? The biologist can no longer turn to nature as a place of grace, as Allee and Emerson once did.

Perhaps fearful that technology will indeed render the natural meaningless and disheartened by the continued environmental destruction of the planet, some have sought a return to organicist metaphors of nature as a path to world salvation. Environmental historian Donald Worster has portrayed the history of ecology as the battle between two approaches to nature. The arcadian image, to which Worster assigns Chicago ecology, is rooted in organicist notions of interdependence, holism, harmony, and balance. This is a romantic vision of nature in which humans, animals, and plants are all a part of the social organism, each respectful of the other in the larger organic whole. Against the arcadian image, Worster places the imperialist tradition, informed by a mechanistic philosophy and tied to technology and the machine; it legitimates, for Worster, a domination of nature rather than a respect. Worster has not, he admits, concealed his "personal conviction" in writing the history of ecology. He believes that "we have had more than enough of imperialism—of that Baconian drive to 'enlarge the boundaries of the human empire.'" "In this age of deadly mushroom clouds and other environmental poisons," he writes, "I believe it is surely time to develop a gentler, more self-effacing ethic toward the earth," an ethic he sees immersed in the arcadian ideal.[32] A similar resurrection of organicist, cooperative metaphors of nature can be found within a certain strand of ecofeminist writings that seeks a return to a value system more in harmony with nature, a value system based in so-called feminine principles of cooperation, nurturance, holism, and interdependence that are seen to arise naturally from a woman's reproductive role as the bearer of children.[33]

Such appeals to nature as cooperative are often founded on liberal renditions of biological determinism. Imparting a particular value system to nature, they eschew the very political foundations that gave substance and definition to the economy of nature. Allee and Emerson, for instance, looked to nature as a justification for value systems that are, first and foremost, political. Yet in asserting through the veil of ecology that nature is cooperative, they closed off debate and thereby obscured the social/political context that gave both definition and meaning to their ecological ideas. Too often we take scientific statements of fact for granted, ignoring their history, ignoring their production. Once produced, such facts become more difficult to challenge, more difficult to question. A natural economy grounded in science dispels oppositional voices, rendering them obsolete and powerless. Yet what is natural for the Amazonian *seringueiros* whose livelihood depends on the extraction of rubber and other resources from the Amazon rain forest, but whose fate is decided meanwhile by international politics and environmental groups from First World countries? One questions whether the scientific view of nature as cooperative and its present-day ties with romanticist notions of primitive nature untouched by human existence are liberating for the *serenguieros*. Susanna Hecht and Alexander Cockburn show, instead, how such romanticist visions of nature sanction the establishment of tropical rain forest preserves that exclude the voice of indigenous tribes.[34] Allee, recall, thought that the one path to world peace entailed the turning of tropical rain forests into high-yielding agricultural areas, which would, theoretically, reduce potential conflicts over shortages of food and other resources.

Definition of nature's economy is, as this study has shown, a contested *political* terrain. In the case of Emerson, for example, the economy of nature implied a hierarchical vision of the natural order that mirrored that of twentieth-century managerial capitalism, where the notion of cooperation was embedded in an organicist view of society that stressed the importance of social control in maintaining the functional efficiency of either the workplace or society as a whole. We must be willing to open discussions about the economy of nature to the political decision-making process, rather than trying to shield such discussions behind the cloak of scientific objectivity. There are no predefined categories, no absolutes to which we can turn, a position once occupied by the idea of the natural, the origin of essence. Where then can we find resolution? Perhaps in our search for identity, meaning, and solace, we should look not to nature but to our own actions. We need to take responsibility for ourselves as a species on this planet, rather than justifying our past decisions or future plans under the guise of "science" and "nature." Nature cannot solve our problems, it cannot give us guidance, it cannot give us direction. Humans, not nature, are the sole arbiters of their fate.

Notes

Readers should consult the Bibliography for full citations of all works mentioned in the notes.

Chapter One

1. Kaye, *Social Meaning of Modern Biology;* and Lewontin, Rose, and Kamin, *Not in Our Genes,* are two works that rather uncritically swallow the straight-line history from social Darwinism to sociobiology. For a defense of the history of naturalistic ethics, see Richards, *Darwin and the Emergence of Evolutionary Theories.* For a sampling of a rather extensive literature on the history of social Darwinism, see Hofstadter, *Social Darwinism;* Bannister, *Social Darwinism;* Bellomy, "'Social Darwinism' Revisited"; and Moore, "Socializing Darwinism." On the naturalization of values, see Young, "Naturalization of Value Systems."

2. Pells, *Liberal Mind in the Conservative Age,* 187.

3. Kaye, *Social Meaning,* 95.

4. The problem has been emphasized by a number of authors. See, e.g., Kitcher, *Vaulting Ambition,* 396–406; Caplan, "Ethics, Evolution, and the Milk of Human Kindness," 304–15.

5. Trivers, "Evolution of Reciprocal Altruism," 35.

6. Donna Haraway has written numerous articles on gender and the body politic metaphor in animal sociology. See, e.g., Haraway, *Primate Visions;* idem, "Animal Sociology"; idem, "Signs of Dominance"; idem, "Teddy Bear Patriarchy." On race and gender metaphors, see Stepan, "Race and Gender." The role of metaphor in feminist critiques of science is also discussed in Keller, *Reflections on Gender and Science;* Harding, *Science Question in Feminism,* 233–39. For a historical analysis of metaphor beyond issues of gender bias, see Young, *Darwin's Metaphor;* and Cantor, "Weighing Light."

7. Arbib and Hesse, *Construction of Reality,* 153, 156, suggest that "almost any interesting descriptive term can be shown etymologically to be a dead metaphor. . . . To make explicit the ramification of metaphor is to engage in critique, evaluation, and perhaps replacement."

Chapter Two

1. For biographical information on Allee, see K. P. Schmidt, "Warder Clyde Allee"; Emerson and Park, "Warder Clyde Allee"; and W. C. Allee, "About Warder Clyde Allee."

2. The Indiana dunes played a central role in the development of ecology at Chicago and in a Midwest movement of social democratic reform. See Engel, *Sacred Sands.*

3. Salisbury and Alden, "Geography of Chicago."

4. Pauly, *Controlling Life,* 67.

5. On the relationship between Chamberlin and Salisbury, see Collie and Densmore, *Thomas C. Chamberlin . . . and Rollin D. Salisbury.* See also R. T. Chamberlin, "Biographical Memoir of Thomas C. Chamberlin"; and *Dictionary of Scientific Biography,* s.v., "Salisbury, Rollin Daniel."

6. Pattison, "Rollin Salisbury and the Establishment of Geography at the University of Chicago," 154. See other essays in this volume for a discussion of various topics pertaining to the professionalization of geography in the United States.

7. Salisbury, "Geology in Education," 334; Salisbury, Barrows, and Tower, *Elements of Geography,* 3; Tower, "Scientific Geography," 806.

8. The constraints of environmental determinism with respect to theories of human progress is discussed in Coleman, "Science and Symbol in the Turner Frontier Hypothesis," 22–49, esp. 32–35.

9. For various accounts of this theory, see Chamberlin and Salisbury, *General Treatise on Geology;* idem, *College Text-book of Geology* (hereafter *College Geology*); R. T. Chamberlin, "Origin and History of the Earth."

10. Chamberlin and Salisbury, *College Geology,* 842–43.

11. Ibid., 915.

12. Ibid., 943.

13. "Geography in the University of Chicago," *Bulletin of the American Geographical Society of New York,* 35 (1903): 207–8. See also Pattison, "Goode's Proposal."

14. For biographical information on Cowles, see Adams and Fuller, "Henry Chandler Cowles"; Cooper, "Henry Chandler Cowles"; and S. G. Cook, "Cowles Bog, Indiana."

15. On the founding of the Hull Biological Laboratories, see Goodspeed, *History of the University of Chicago,* 301–7.

16. On Coulter, see A. D. Rodgers, *John Merle Coulter.*

17. Cittadino, "Ecology and the Professionalization of Botany in America, 1890–1905."

18. Coulter, *Plant Relations.*

19. See Coleman, "Evolution into Ecology?" for a discussion of Warming's ecological views.

20. Cowles, "Ecological Relations of the Vegetation." For a discussion of the influence of Cowles on the development of American plant ecology, see

McIntosh, *Background of Ecology,* 39–49; Tobey, *Saving the Prairies,* 106–9; Worster, *Nature's Economy,* 206–8.

21. Cowles, "Ecological Relations," 95.

22. Cowles, "Physiographic Ecology of Chicago and Vicinity," 74. This article was also published as "Plant Societies of Chicago and Vicinity." Engel, in *Sacred Sands,* 135–68 discusses the shared physiographic context of Cowles's and Salisbury's work on the Chicago region. For examples of the effect of the formation concept on Nebraskan ecology, see Tobey, *Saving the Prairies,* 60–70, 87–99; and Hagen, "Ecologists and Taxonomists."

23. Cowles, "Plant Societies," 8. On Cowles's later modification of his views, see his "Causes of Vegetative Cycles."

24. *Annual Register, 1897–1898* (Chicago: University of Chicago Press, 1898), 322.

25. Hagen, "Organism and Environment," 272.

26. N. Taylor, "Some Modern Trends in Ecology," 113.

27. Coleman, "Science and Symbol."

28. Chamberlin and Salisbury, *College Geology,* 942–3.

29. Goode, "Human Response to the Physical Environment," 343.

30. Salisbury to Davenport, Davenport to Salisbury, and Davenport to Harper, all dated 16 November 1901, box 18, folder 7, University of Chicago Presidents' Papers, 1889–1925, Joseph Regenstein Library, University of Chicago (hereafter UCPP, 1889–1925).

31. For biographical information on Whitman, see Lillie, "Charles Otis Whitman"; Maienschein, "Introduction"; Morse, "Charles Otis Whitman."

32. For a history of Clark University and its importance for the University of Chicago, see Ryan, *Studies in Early Graduate Education,* 47–63, 117–21; Storr, *Harper's University,* 65–77.

33. Blake, in "Concept and Development of Science at the University of Chicago," provides an informative discussion of the interactions between Whitman and Harper and the formation of the biological division at Chicago. Whitman's plan for the biological sciences is also discussed in Pauly, *Controlling Life,* 58–60, 64–65; and Maienschein, "Whitman at Chicago," 151–82.

34. Whitman, "Specialization and Organization."

35. Whitman, "Biological Instruction in the Universities," 513. Both Maienschein, in "Whitman at Chicago," and Pauly, in "Summer Resort and Scientific Discipline," emphasize this organizational ideal in Whitman's work.

36. The pivotal role the MBL played in the fostering and direction of biological research in America is discussed in Lillie, *Woods Hole Marine Biological Laboratory;* Maienschein, "Introduction"; Manning, *Black Apollo of Science,* 67–114; and Pauly, "Summer Resort and Scientific Discipline."

37. "Doctors of Philosophy (June 1893–April 1931)," *Announcements: University of Chicago* (1931), 34–35.

38. Whitman, "Inadequacy of the Cell-Theory of Development," 646. For a historical analysis of Whitman's early embryological research, see Maienschein, "Cell Lineage."

39. Some of this material was published in Carr, ed., *Posthumous Works of Charles Otis Whitman.*

40. Whitman, "Animal Behavior," 328, 299–300.

41. For a discussion of Whitman's orthogenetic views, see Bowler, *Eclipse of Darwinism,* 159–60.

42. Whitman to Yerkes, 24 October 1909, II A, folder 99, Frank R. Lillie Papers, Library of the Marine Biological Laboratory, Woods Hole, Mass. (hereafter FRLP-MBL).

43. Burkhardt, "Charles Otis Whitman, Wallace Craig, and the Biological Study of Animal Behavior."

44. Craig, "Expression of the Emotions in the Pigeons"; Turner, "The Homing of Ants." Wheeler received his Ph.D. under Whitman at Clark University in the spring of 1892 for research on insect embryology. Wheeler migrated to the University of Chicago with Whitman, serving on the faculty from 1892 to 1899. Wheeler did not begin his distinguished career researching the behavior, taxonomy, and natural history of the ant until he left Chicago. On Wheeler, see M. A. Evans and H. E. Evans, *William Morton Wheeler, Biologist.* Samuel Jackson Holmes's dissertation also focused on problems of development, although he later wrote a number of treatises on animal behavior. See S. J. Holmes, *Evolution of Animal Intelligence;* and idem, *Studies in Animal Behavior.*

45. Loeb, *Mechanistic Conception of Life,* 41.

46. Whitman, "Some of the Functions and Features of a Biological Station."

47. Maienschein, in "Physiology, Biology, and the Advent of Physiological Morphology," points out the appeal of Loeb's research in physiological morphology to the MBL crowd.

48. Pauly, *Controlling Life,* 92.

49. Biographical information on Child is available in Hyman, "Charles Manning Child." For an analysis of his regeneration studies in light of his physiological theory of inheritance, see Mitman and Fausto-Sterling, "Whatever Happened to *Planaria?*"

50. Child, *Physiological Foundations of Behavior,* 21.

51. Ibid., 5.

52. Child, "Some Considerations Regarding So-Called Formative Substances," 180.

53. Ibid., 170.

54. Child, "Relation between Functional Regulation and Form Regulation," 580.

55. Child, "Behavior Origins from a Physiologic Point of View," 182.

56. Child, *Physiological Foundations of Behavior,* 30. On the relations between Child's ideas and those of Chicago pragmatists, see Kingsland, "Toward a Natural History of the Human Psyche." For a discussion of Chicago pragmatism, see Rucker, *Chicago Pragmatists;* Marcell, *Progress and Pragmatism;* Lewis and Smith, *American Sociology and Pragmatism.*

57. Dewey, "Reflex Arc Concept."

58. Rucker, *Chicago Pragmatists,* 94.

59. Dewey, *Democracy and Education,* 45.

60. For biographical information on Jennings, see Sonneborn, "Herbert Spencer Jennings." For a discussion of the Loeb-Jennings debate, see Pauly, *Controlling Life,* 118–29, and idem, "Loeb-Jennings Debate."

61. Jennings, *Behavior of the Lower Organisms,* 270.

62. Ibid., 313.

63. For historical analyses of this separation, see Gilbert, "Embryological Origins of the Gene Theory"; idem, "Cellular Politics"; Sapp, *Beyond the Gene.*

64. Johannsen, "Genotype Conception of Heredity," 159.

65. Child, "Process of Reproduction in Organisms," 36.

66. Child, *Physiological Foundations of Behavior,* iii.

67. For biographical information on Lillie, see Willier, "Frank Rattray Lillie."

68. Lillie, "Theory of Individual Development," 245.

69. Ibid., 244.

70. Child to Lillie, 8 August 1924, box 2, folder 13, Frank Rattray Lillie Papers, Joseph Regenstein Library, University of Chicago (hereafter FRLP-UC).

71. Whitman to Harper, 12 September 1899, folder 6, box 18, UCPP, 1889–1925.

72. Davenport to Allee, 20 December 1938, Box 7, Folder 3, Warder Clyde Allee Papers, Joseph Regenstein Library, University of Chicago (hereafter WCAP). On Davenport's experimental morphology, see his *Experimental Morphology.*

73. On these various excursions, see *Botanical Gazette* 30 (1900): 142; *Botanical Gazette* 33 (1902): 400.

74. Davenport, "Animal Ecology of the Cold Spring Sand Spit," 172.

75. Davenport, "The Collembola of Cold Spring Beach," 24.

76. Davenport, "Animal Ecology of the Cold Spring Sand Spit," 173.

77. On the theory of organic selection, see Richards, *Darwin and the Emergence of Evolutionary Theories,* 469–74, 480–95.

78. Quoted in Kevles, *In the Name of Eugenics,* 45.

79. *Annual Register, 1901–1902,* (Chicago: University of Chicago Press, 1902), 311.

80. For biographical information on Adams, see Palmer, "Resolution of Respect, Dr. Charles C. Adams; Sears, "Charles C. Adams, Ecologist."

81. Adams, "Southeastern United States as a Center of Geographical Distribution of Flora and Fauna," 126.

82. Adams, "Postglacial Dispersal of the North American Biota," 68. For other relevant papers, see idem, "Baseleveling and Its Faunal Significance"; idem, "Variations and Ecological Distribution of the Snails of the Genus Io."

83. Shelford, "Physiological Animal Geography," 613. For biographical information on Shelford, see Kendeigh, "Victor Ernest Shelford"; *Who Was Who in America,* s.v. "Shelford, Victor Ernest."

84. Shelford's published dissertation included "Preliminary Note on the

Distribution of the Tiger Beetles" and "Life Histories and Larval Habits of the Tiger Beetles."

85. *Annual Register (1907–1908)* (Chicago: University of Chicago Press, 1908), 352.

86. Shelford, "Physiological Animal Geography," 553.

87. Shelford, "Ecological Succession, V," 333–34.

88. Shelford, "Comparison of the Responses of Sessile and Motile Plants and Animals," 661, 665.

89. Ibid., 665. For an institutional analysis of the relationship between the development of biology and medicine in the United States, see Pauly, "Appearance of Academic Biology in Late Nineteenth-Century America." On the development of general physiology, see idem, "General Physiology and the Discipline of Physiology."

90. Official academic record of Victor Ernest Shelford, University of Chicago, Office of the Registrar; Shelford, "Physiological Animal Geography," 554.

91. Shelford, "Physiological Animal Geography," 594.

92. These categories are found in both Adams, *Guide to the Study of Animal Ecology;* and Shelford, "Principles and Problems of Ecology."

93. Official academic record of Warder Clyde Allee, University of Chicago, Office of the Registrar.

94. See W. C. Allee, "Experimental Analysis of the Relation between Physiological States and Rheotaxis in Isopoda."

95. Shelford and W. C. Allee, "Reaction of Fishes to Gradients and Dissolved Atmospheric Gases," 261; idem, "Index of Fish Environments"; idem, "Rapid Modification of Behavior of Fishes by Contact with Modified Water"; W. C. Allee, "Further Studies on Physiological States and Rheotaxis in Isopoda"; idem, "Ecological Importance of the Rheotactic Reaction of Isopods"; Wells, "Reactions and Resistance of Fishes in Their Natural Environment to Salts"; Phipps, "Experimental Study of the Behavior of Amphipods."

96. On Craig, see Burkhardt, "Charles Otis Whitman, Wallace Craig."

97. Craig, "North Dakota Life," 412.

98. Shelford, *Animal Communities in Temperate America,* 35.

99. Shelford, "Experimental Study of the Behavior Agreement."

100. Shelford, "Ecological Succession, I–V."

101. Shelford, "Ecological Succession, I," 28.

102. Barrows, "Geography as Human Ecology," 3.

103. Lillie to Allee, 12 November 1920, box 1, folder 3, FRLP-UC.

Chapter Three

1. W. C. Allee, "About Warder Clyde Allee," 229. For an autobiographical account of Allee's experiences and impression of the MBL, see untitled manuscript, Allee correspondence, 1905–38, Edwin M. Banks Collection, University of Illinois Archives, Urbana (hereafter EBC). Allee's

community studies at Woods Hole were published as a series of papers. See W. C. Allee, "Studies in Marine Ecology," I, III, IV.

2. M. H. Allee, "Recipe for a Book," 7. On Hill Allee's life and career, see Winslow, "Marjorie Hill Allee."

3. M. H. Allee, *Jane's Island,* 27, 53.

4. Ibid., 55–56.

5. W. C. Allee, *Animal Aggregations,* 3.

6. W. C. Allee, "Science Confirms an Old Faith," 780.

7. M. H. Allee, *Jane's Island,* 139–40.

8. W. C. Allee, "Evolution of a Mechanist," ca. 1915, Allee correspondence, 1905–38, EBC, 9.

9. Weinstein in "A Note on W. L. Tower's *Leptinotarsa* Work," 352–53 suggests that Tower resigned from the department in 1915 as the result of suspicions that his work on the evolution and inheritance of potato beetles was faked. It is clear, however, from newspaper reports that Tower did not resign until 1917, and the immediate cause was a highly contested divorce. See "Prof. Tower and Wife Both Race For Freedom," *Chicago Tribune,* 5 May 1917.

10. W. C. Allee, "Evolution of a Mechanist," 9–11.

11. Untitled manuscript ca. 1916, Allee correspondence, 1905–38, EBC, 25.

12. Quoted in G. Jones, *On Doing Good,* 14.

13. Chatfield, "World War I and the Liberal Pacifist in the United States," 1921.

14. Ibid., 1920.

15. Flitcraft, ed., *History of the 57th Street Meeting of Friends;* "Protest Abuse of Quakers," *Chicago Tribune,* 2 December 1916.

16. "What Happens in Military Prisons: The Public Is Entitled to the Facts," Allee Research Correspondence, 1983–84, EBC. For a discussion of the activities of the American Friends Service Committee, see Forbes, *The Quaker Star under Seven Flags;* R. M. Jones, *A Service of Love in Wartime;* L. M. Jones, *Quakers in Action.*

17. Allee to Addams, 19 March 1917, Woman's Peace Party Papers, Swarthmore College Peace Collection, Swarthmore, Pa.; Addams to Allee, 3 June 1924, box 10, folder 1, WCAP; W. C. Allee, "About Warder Clyde Allee"; M. H. Allee, "History of the Chicago Meeting of Orthodox Friends Compiled from Monthly Meeting Minutes, First Month 15th, 1913 to Eighth Month 17th, 1921," 57th Street Meeting of Friends, Chicago.

18. This shift in the American peace movement toward the political left during World War I is discussed in Chatfield, "World War I"; idem, *For Peace and Justice;* DeBenedetti, *Origins of the Modern American Peace Movement;* idem, ed., *Peace Heroes;* C. Roland Marchand, *The American Peace Movement.*

19. Chatfield, "World War I," 1936.

20. DeBenedetti, *Peace Heroes,* 5.

21. Molly Bath to author, 12 November 1989. Allee's voting record is

also evident in a number of letters. See Allee to Wright, Election Day 1936, box 24, folder 11; and Allee to mother, 1 June 1952, box 10, folders 4–7, WCAP. For Allee's opposition to the New Deal, see W. C. Allee, "Biology of Democracy," Earlham College Commencement Address 1934, EBC.

22. Quoted in Chatfield, *For Peace and Justice,* 62.

23. Thomas, *Conscientious Objector in America,* 29.

24. Addams, *Peace and Bread,* 134–35.

25. "City Pacifists Ask Wilson for War Referendum," *Chicago Tribune,* 7 February 1917.

26. *Springfield News-Record,* 9 February 1917. See, also, "Why Is a Newspaper," *Stentor* (1916–17):144 and "An Important Correction," ibid., 141 (Lake Forest, Ill.: Lake Forest College).

27. W. C. and M. H. Allee, "Adventures of a College Professor," 1133.

28. "German Teacher at Lake Forest Seized from Mob," *Chicago Tribune,* 8 April 1917. On Schmidt's dismissal, see Faculty Minute Papers, Lake Forest College.

29. For biographical information on Schmidt, see *Copeia* 3 (1959): 189–92; Emerson, "K. P. Schmidt". On the Schmidt-Needham relationship, see Schmidt-Needham correspondence, box 14, folder 1, Karl Patterson Schmidt Papers, Field Museum of Natural History Library, Chicago (hereafter KPSP).

30. Needham to Schmidt, 29 May 1917, box 14, folder 1, KPSP.

31. For biographical information on Emerson, see E. O. Wilson and Michener, "Alfred Edwards Emerson"; and T. Park, "Alfred Edwards Emerson."

32. Emerson to Schmidt, 28 May 1917, 17 November 1917, 15 April 1918, box 6, folder 4, KPSP.

33. For a discussion of the National Research Council's formation see Daniel J. Kevles, "George Ellery Hale"; idem, *The Physicists,* 102–16; Yerkes, ed., *New World of Science.*

34. Bigelow, "Contributions of Zoology to Human Welfare," 3.

35. For a discussion of the relationship between German military philosophy and Darwinism, see Daniel Gasman, *Scientific Origins of National Socialism.* Gasman argues that Haeckel provided the Pangerman League with scientific credibility for its social Darwinism and racial elements. Kelly, in *The Descent of Darwin,* 100–122, argues, however, that the tendency to see social Darwinism as the historical precursor to World War I nationalism and later Nazism is misleading.

36. The number of American biologists who addressed the supposed connection between Darwinism and German militarism is substantial. See Bigelow, "Contributions of Zoology"; Cole, "Biological Philosophy and the War"; Jordan, "Social Darwinism"; idem, *Root of Evil;* idem, *Democracy and World Relations,* 51–55; Kellogg, "War and Human Evolution"; idem, "War for Evolution's Sake"; idem, *Headquarters Nights;* Loeb, "Biology and War"; Menge, "Darwinism, Militarism, Socialism"; Metcalf, "Darwinism and Nations"; Osburn, "Some Common Misconceptions of Evolution"; Patten,

"Message of the Biologist"; Pearl, "Biology and War"; Ritter, "Business Man's Appraisement of Biology."

37. Mitchell's *Evolution and the War* was the first major work to address specifically the relationship between German military ideology and Darwinism. Mitchell's work, however, had a relatively minor influence in America compared to *Headquarters Nights,* perhaps because Mitchell refused to entertain determinist arguments. Of all the papers by American biologists discussing German war ideology and Darwinism, only Menge and Osburn cite Mitchell's work. For a thorough analysis of Mitchell's work and antiwar biology in Britain, see D. P. Crook, "Peter Chalmers Mitchell."

38. Biographical information on Kellogg is available in McClung, "Vernon Lyman Kellogg," and *National Cyclopedia of American Biography,* s.v. "Kellogg, Vernon L." For an account of the close relationship between Kellogg and Jordan, see Jordan, *Days of a Man.*

39. Kellogg, *Headquarters Nights,* 23.

40. Ibid., 22.

41. Ibid., 13.

42. Ibid., 28.

43. Ibid., 29–30.

44. Philip Pauly, "Summer Resort," 140; Kelly, *Descent of Darwin,* 100–122.

45. In a similar fashion, the social science professions used their contributions to wartime propaganda to strengthen their own importance within American society. See Gruber, *Mars and Minerva.*

46. For a discussion of Weismann's views, see Churchill, "August Weismann and a Break from Tradition."

47. Kellogg's *Darwinism Today* is the major primary source for a discussion of the issues surrounding Darwinian evolution in the early twentieth century. Bowler, *Eclipse of Darwinism,* is the only secondary work devoted exclusively to a discussion of anti-Darwinian theories during this period. There is, however, a fairly large body of literature that focuses on Mendelism, mutationism, and Weismann. See, e.g., Bowler, "Hugo De Vries and Thomas Hunt Morgan"; Allen, "Thomas Hunt Morgan and the Problem of Natural Selection"; idem, *Life Science in the Twentieth Century,* 41–72, 126–45; Darden, "William Bateson and the Promise of Mendelism"; and Provine, *Origins of Theoretical Population Genetics.*

48. Pearl, "Biology and War," 355. See also Kellogg, "War for Evolution's Sake," 153–55.

49. Jordan, *Days of a Man,* 2:735. For a biographical account that focuses on Jordan's politics and his career as Stanford's first president, see Burns, *David Starr Jordan.* Spoehr, "Progress' Pilgrim," provides a detailed analysis of Jordan's ideas in the context of progressive reform.

50. Little attention has been given to the peace activist side of eugenics in the secondary literature. See Haller, *Eugenics,* 87–88, for a brief discussion of Jordan's views. For a brief analysis of British peace eugenists, see Stepan, "'Nature's Pruning Hook.'"

51. See, e.g., Jordan, *Human Harvest;* idem, *War and the Breed;* idem, *War's Aftermath;* idem, "War and Genetic Values"; Kellogg, "Eugenics and Militarism"; idem, "Bionomics of War; idem, *Military Selection and Race Deterioration.*

52. Kellogg, "Eugenics and Militarism," 101.

53. Kellogg, "The Bionomics of War," 44–52.

54. Jordan, *War and the Breed.*

55. Jordan, "Social Darwinism," 401.

56. Spencer, *Principles of Sociology,* 592.

57. Richards, *Darwin and the Emergence of Evolutionary Theories,* 261.

58. Darwin, *Origin of Species,* 62.

59. Richards, *Darwin,* 211.

60. Espinas, *Sociétés animales,* 153. On the connections between Spencer and Espinas, see Richards, *Darwin,* 329.

61. Espinas, *Sociétés animales,* 546. On Espinas's writings in the context of individualism and nationalism, see Guillame, "La Biosociologie d'Espinas."

62. For an autobiographical account of Kropotkin's life, see Kropotkin, *Memoirs.* The most comprehensive historical account of Kropotkin's ideas, which places them within the context of Russian evolutionary thought, is Todes, *Darwin without Malthus.*

63. On Kropotkin's Lamarckian position, see Kropotkin, "Direct Action of the Environment on Plants"; idem, "Inheritance of Acquired Characters"; idem, "Direct Action of Environment and Evolution."

64. On Mitchell and the importance of ecological, cooperationist views in British antiwar evolutionism, see Crook, "Peter Chalmers Mitchell"; and idem, "Nature's Pruning Hook?"

65. Shelford, *Animal Communities,* 8.

66. Mitchell, *Evolution and the War,* 41.

67. See Numbers, "Creationism in Twentieth-Century America."

68. Marsden, in *Fundamentalism and American Culture,* places particular importance on World War I with respect to the emergence of the fundamentalist movement and its antievolution campaign in the 1920s. See also Gatewood, *Controversy in the Twenties;* Levine, *Defender of the Faith;* Moore, *The Post-Darwinian Controversies.*

69. Marsden, *Fundamentalism,* 161.

70. See, e.g., Conklin, "Biology and Democracy"; idem, *Direction of Human Evolution;* Jordan, *Ways to Lasting Peace;* idem, *Democracy and World Relations;* idem, "Search for the Master Key"; Patten, "Message of the Biologist"; idem, *Grand Strategy of Evolution;* Ritter, *War, Science and Civilization;* idem, "Business Man's Appraisement of Biology"; idem, "Biology's Contribution to a System of Morals"; idem, *Unity of the Organism.*

71. Patten, *Evolution,* 163.

72. Patten, *Growth,* 6–8. For a more in-depth historical analysis of

Patten in the context of biology and the First World War, see Mitman, "Evolution as Gospel."

73. Conklin, "Biology and Democracy," 407.

74. See, e.g., Filene, "Obituary for 'The Progressive Movement.'"

75. D. Rodgers, "In Search of Progressivism." The link between certain aspects of progressivist thought and the Protestant social gospel is most systematically explored in Crunden, *Ministers of Reform.* See also Crunden, in *Progressivism,* 71–104; Engel, *Sacred Sands,* 45–79; Griffen, "Progressive Ethos"; and Higham, "Hanging Together." For other works that emphasize the notion of community and the stress placed on organic society in progressivist thought, see Crunden, *From Self to Society;* Pells, *Radical Visions and American Dreams;* Price, "Community and Control"; Quandt, *From the Small Town to the Great Community;* and Violas, "Progressive Social Philosophy."

76. Dewey, "Relation of Philosophy to Theology," 367.

77. The role of social Darwinism in late nineteenth- and early twentieth-century American social thought is the subject of an ongoing and lengthy controversy. The classic study is Hofstadter, *Social Darwinism.* Hofstadter argues that social Darwinism was a pervasive theme in late nineteenth-century America. Wyllie, in "Social Darwinism and the American Businessman," argues that there is little evidence to suggest the use of social Darwinist rhetoric among the American business community. Bannister, in *Social Darwinism,* suggests, in direct opposition to Hofstadter, that social Darwinism was, in fact, a myth created by American reformers and later by historians such as Hofstadter to serve as a springboard for their own liberal ideas. For two recent critical surveys, see Bellomy, "'Social Darwinism' Revisited"; and Moore, "Socializing Darwinism."

78. Crunden, *Ministers of Reform,* 65.

79. Allen, "Eugenics Record Office at Cold Spring Harbor," 258. See also Kevles, *In the Name of Eugenics;* and Pickens, *Eugenics and the Progressives.*

80. Patten, "Message of the Biologist," 101.

81. Patten, *Evolution,* 174.

Chapter Four

1. See, e.g., Pearse, *Animal Ecology;* and Elton, *Animal Ecology.*

2. For an analysis of the historical development of theoretical population ecology, see Kingsland, *Modeling Nature.* See also T. Park, "Some Observations on the History and Scope of Population Ecology"; and McIntosh, *Background of Ecology,* 146–92. On theoretical population genetics, see Provine, *Origins of Theoretical Population Genetics;* and idem, *Sewall Wright and Evolutionary Biology.*

3. See, e.g., W. C. Allee, "Further Studies on Physiological States and Rheotaxis in Isopoda"; idem, "Certain Relations between Rheotaxis and Resistance to Potassium Cyanide in Isopoda"; idem, "Chemical Control of Rheo-

taxis in Asellus"; and W. C. Allee and Stein, Jr., "Light Reactions and Metabolism in May-Fly Nymphs."

4. For a popular account of the Allees' experiences in Barro Colorado, see W. C. Allee and M. H. Allee, *Jungle Island.* See also W. C. Allee, "Distribution of Animals in a Tropical Rain-Forest"; idem, "Measurement of Environmental Factors in the Tropical Rain-Forest of Panama."

5. W. C. Allee, *Animal Life and Social Growth,* 99.

6. For other historical studies of Allee's professional career, see Banks, "Warder Clyde Allee and the Chicago School of Animal Behavior"; Caron, "La Théorie de la coopération animale dans l'écologie de W. C. Allee."

7. W. C. Allee, *Animal Aggregations,* 9.

8. Allee to Phillips, 13 February 1933, box 21, folder 3, WCAP.

9. Reuter, "Relation of Biology and Sociology," 712.

10. Wheeler, "Animal Societies," 290. For biographical information on Wheeler, see M. A. Evans and H. E. Evans, *William Morton Wheeler.*

11. Braun-Blanquet, *Plant Sociology.*

12. W. C. Allee, "Animal Aggregations: A Request for Information," 129–30.

13. Allee's research notebooks on these early experiments are in box 1, folder 8, WCAP. The results of these initial experiments were first presented at the annual meeting of the American Society of Zoologists in December 1919. See W. C. Allee, "Animal Aggregations."

14. For a comprehensive analysis of the nominalist/realist debate in American sociology and in the Chicago school of sociology in particular, see Lewis and Smith, *American Sociology and Pragmatism.*

15. W. C. Allee, *Animal Aggregations,* 51.

16. W. C. Allee to Collias, 14 April 1942, Allee correspondence 1942–55, EBC. The article referred to is Tarde, "Inter-psychology." For a discussion of Tarde's influence on sociological thought, see Clark, ed., *Gabriel Tarde,* 1–72.

17. W. C. Allee, "Cooperation among Animals," 423.

18. See, e. g., W. C. Allee, "Effect of Bunching on Oxygen Consumption in Land Isopods"; idem, "Studies in Animal Aggregations: Causes and Effects of Bunching"; idem, "Studies in Animal Aggregations: Some Physiological Effects."

19. W. C. Allee, "Animal Aggregations," 375.

20. These studies include Bohn and Drzewina, "Variations de la sensibilité"; Drzewina and Bohn, "La Défense des animaux"; idem, "Variations de la susceptibilité aux agents nocifs"; idem, "Sur des phénomènes d'auto-protection"; idem, "Action nocive de l'eau"; idem, "Variations dans le temps de la résistance"; idem, "Sur des Phénomènes d'auto-destruction."

21. See the comments of Pieron and Lapicque in Drzewina and Bohn, "Action nocive."

22. W. C. Allee and Schuett, "Studies in Animal Aggregations," 312. In this respect, Allee agreed with the findings of Bresslau whose work on protozoa demonstrated the secretion of a substance he called tektin that absorbed toxic reagants. See Bresslau, "Die Ausscheidung."

23. Oesting and W. C. Allee, "Further Analysis of the Protective Value of Biologically Conditioned Fresh Water," 325. See also W. C. Allee, "Studies in Animal Aggregations: Mass Protection From Hypotonic Sea-Water"; idem, "Studies in Animal Aggregations: Mass Protection From Fresh Water."

24. Shelford, *Animal Communities;* and Elton, *Animal Ecology.* For a discussion of Elton's importance in the development of animal ecology, see Cox, "Charles Elton and the Emergence of Modern Ecology."

25. Allee and Schuett, "Studies in Animal Aggregations," 314–15.

26. Ibid., 315.

27. Allee, *Animal Life and Social Growth,* 149. Allee's first statements regarding the principle of cooperation appeared in his "Studies in Animal Aggregations: Mass Protection from Fresh Water"; idem, "Science Confirms an Old Faith"; and idem, "Cooperation among Animals."

28. Pearl, "Evolution of Sociality," 306.

29. The fact that Allee thought sociality arose from the beneficial survival value of groups is further supported by Peter Frank's recollections of his days as a graduate student under Allee. Frank remarked, "I recall that I was struck, in 1952, by G. E. Hutchinson's statement that true sociality had certainly arisen from family relations rather than the subsocial phenomena Allee dealt with. That I had not been introduced to this idea explicitly is, I think, symptomatic of Allee's thoughts on the subject," Frank to Banks, 27 May 1983, letters regarding Allee 1983–85, EBC.

30. Allee, *Animal Aggregations,* 355.

31. W. C. Allee and Rosenthal, "Group Survival Value for *Philodina roseola.*"

32. Allee, *Animal Aggregations,* 47. See also Deegener, *Die Formen der Vergesellschaftung in Tierreiche.*

33. Wheeler, "Societal Evolution," 145. See also Wheeler's *Social Life among the Insects, Emergent Evolution and the Development of Societies,* and *Social Insects* for a discussion of the importance of the family in social evolution.

34. G. S. Miller, Jr., "Some Elements of Sexual Behavior in Primates," 278; Zuckerman, *Social Life of Monkeys and Apes.*

35. Haraway, in *Primate Visions,* provides a provocative cultural analysis of primatology. The first section of her book emphasizes the extent to which organic functionalism, with its implicit assumptions of male dominance, hierarchy, and division of labor, echoed across the biological, social, and political landscape in the pre–World War II era.

36. On Linton, see A. Linton and Wagley, *Ralph Linton.*

37. In a letter to me, Molly Barth, Allee's youngest daughter, mentioned Ralph Linton as one of her father's few interdisciplinary friendships that she could recall. Barth to Mitman, 12 November 1989.

38. See Schmidt to Allee, 10 November 1925; Allee to Schmidt, 8 July 1926, box 1, folder 3, KPSP.

39. Hesse, *Ecological Animal Geography.*

40. See Allee to Osgood, 6 December 1931, box 1, folder 3, KPSP.

41. Flitcraft, *History of 57th Street Meeting of Friends.*

42. R. Linton, *Study of Man,* 209.

43. Malinowski, *Sex and Repression in Savage Society,* 184; R. Linton, *Study of Man,* 209.

44. R. Linton, *Study of Man,* 92, 150.

45. Ibid., 151, 220.

46. Ibid., 92, 108.

47. Ibid., 93.

48. Ibid., 94.

49. Child, *Physiological Foundations of Behavior,* 271.

50. Ibid., 288.

51. Ibid., 297.

Chapter Five

1. For biographical information on Pearl, see Allen, "Old Wine in New Bottles"; Jennings, "Raymond Pearl"; Kingsland, *Modeling Nature,* 56–76.

2. Pearl, *Biology of Population Growth,* 2.

3. For a discussion of the Immigration Restriction Act in the context of eugenics, see Allen, "Eugenics Record Office"; Kevles, *In the Name of Eugenics,* 96–97. For a history of eugenics in America, see Mark Haller, *Eugenics;* Ludmerer, *Genetics and American Society;* Mehler, "History of the American Eugenics Society"; Pickens, *Eugenics and the Progressives.*

4. On the connections between eugenics and birth control, see Gordon, "Politics of Population"; and idem, *Woman's Body, Woman's Right.*

5. Dublin, ed., *Population Problems.*

6. On the history of theoretical population ecology, see Kingsland, *Modeling Nature.*

7. W. C. Allee, "Biology of Disarmament"; idem, "Biology of Peace."

8. On the history of the LCRC, see Bulmer, "Early Institutional Establishment of Social Science Research"; idem, *Chicago School of Sociology;* T. V. Smith and White, eds., *Chicago.*

9. There is little evidence in the archival records to suggest that there was any interaction between Allee and Park. Thomas Park, a graduate student of Allee's during the early 1930s and later a colleague, did not recall any associations with the sociology department (author's interview, 23 July 1985). To my knowledge, Robert Park cited Allee's work only once. See R. Park, *Human Communities,* 252, 259. For a discussion of Park's ideas and his influence on American sociology, see Bulmer, *Chicago School of Sociology;* Kurtz, *Evaluating Chicago Sociology;* Matthews, *Quest for an American Sociology.*

10. R. Park, "Human Nature and Collective Behavior," 735.

11. For a discussion of the influence of Child on Park, see Bulmer, *Chicago School,* 115; and Hughes, "On Becoming a Sociologist."

12. R. Park, *Human Communities,* 152, 162.

13. Allee to Merriam, 26 October 1927; see also Allee to Merriam, 22 July 1927; Merriam to Allee, 27 July 1927; Allee to Merriam, 18 October

1927; Merriam to Allee, 20 October 1927, box 24, folder 19, Charles E. Merriam Papers, Joseph Regenstein Library, University of Chicago (hereafter CEMP).

14. Merriam to Mason, 12 January 1927; "Memorandum on additional personnel for the development of social science," box 122, folder 9, CEMP.

15. On the committee's interest in Carr-Saunders, see "Report of Progress on Four Year Research Program," box 112, folder 6; and Marshall to Merriam, 3 January 1928, box 122, folder 7, CEMP. For biographical information on Carr-Saunders, see *Dictionary of National Biography,* s.v. "Carr-Saunders, Alexander."

16. Carr-Saunders, *Population Problem,* 197–242.

17. Wynne-Edwards, *Animal Dispersion.* Gilpin *(Group Selection),* J. M. Smith ("Group Selection"), and Sober *(Nature of Selection)* all indicate that Carr-Saunders's ideas anticipated those of Wynne-Edwards without exploring the historical connections. Although Wynne-Edwards claims to have read Carr-Saunders only after writing *Animal Dispersion,* both Pearl and Allee were familiar with Carr-Saunders's work, as were other population theorists and ecologists of the 1930s.

18. Allen, "Old Wine in New Bottles," 3.

19. On the funding of Pearl's institute, see ibid.; and Kingsland, *Modeling Nature,* 61–63.

20. For a history of the NRC-CRPS and a breakdown of the funds received, see Aberle and Corner, *Twenty-Five Years of Sex Research.* On the emergence of reproductive science in the United States, see Clarke, "Emergence of the Reproductive Research Enterprise."

21. Newman to Lillie, 10 September 1912, IIA 61, FRLP-MBL. For a history of the Chicago Department of Zoology, see Newman, "History of the Department of Zoology." On Lillie and Newman's family relationship, see Newman to Lillie, 13 March 1911, IIA 100.58, and Lillie to Newman, 17 March 1911, IIA 100.59, FRLP-MBL. Tower's dismissal was covered in the *Chicago Tribune.* See "Prof. Tower and Wife Both Race for Freedom," *Chicago Tribune,* 5 May 1917.

22. On Lillie's initial theory of the freemartin, see his "Free-martin." On the funding relationships between Chicago and the NRC-CRPS, including a full bibliography of Chicago papers published on sex biology, see Aberle and Corner, *Twenty-Five Years of Sex Research.* Information on Lillie's grants from the NRC-CRPS is also available in RG1.1 200, Box 40, folder 453, Rockefeller Foundation Archives, Rockefeller Archive Center, North Tarrytown, N.Y. (hereafter RFA). On the importance of research materials in reproductive biology, see Clarke, "Research Materials." For historical and sociological analyses of sex hormone research and reproductive physiology, see idem, "Emergence of the Reproductive Research Enterprise"; Hall, "Biology, Sex Hormones and Sexism"; Oudshoorn, "Endocrinologists and the Conceptualization of Sex"; idem, "On the Making of Sex Hormones."

23. On Lillie's membership in these eugenic societies, see Blount to Lillie, 19 May 1921, box 3, folder 8, FRLP-UC.

24. Lillie, "Department of Zoology in Relation to the New Organization," 11 December 1930, box 19, folder 14–15, WCAP.

25. Lillie to Rose, 17 January 1924, RG1.1 216D, box 8, folder 104, RFA. Lillie, "Department of Zoology," WCAP.

26. Lillie to Rose, 17 June 1924, RG1.1 216D, box 8, folder 104, RFA.

27. Lillie to Hutchins, 23 June 1930, box 109, folder 2, UCPP, 1925–45. For an institutional analysis of the development of biology programs and their relationship to medical education, see Pauly, "Appearance of Academic Biology."

28. Lillie to Hutchins, 23 June 1930, box 109, folder 2, UCPP, 1925–45; Lillie to Mason, 5 June 1931, RG 1.1 216D, box 8, folder 104, RFA.

29. Lillie to Hutchins, 23 June 1930, UCPP. On the tension between embryologists and geneticists during this period, see Gilbert, "Cellular Politics"; and Sapp, *Beyond the Gene*.

30. Lillie to Hutchins, 23 June 1930, UCPP.

31. On Newman's course offerings, see the *Annual Register* between 1911 and 1940. Newman's twin studies are discussed in Kevles, *In the Name of Eugenics,* 141–42. See also Newman, Freeman, and Holzinger, *Twins*.

32. Child to Lillie, 8 August 1924, box 2, folder 13, FRLP-UC.

33. Allee to Lillie, undated (ca. 1924), box 19, folder 14–15, WCAP. Child to Lillie, 8 August 1924, box 2, folder 13, FRLP-UC. For a biographical study of Wright and his contributions to population genetics and evolutionary biology, see Provine, *Sewall Wright*.

34. Lillie to Hutchins, 23 June 1930, UCPP.

35. On Wright's opinion of eugenics, see Provine, *Sewall Wright,* 110, 180–82.

36. Lillie to Judson, 7 February 1917, box 18, folder 7, UCPP, 1889–1925.

37. Allee to Lillie, 14 November 1926, box 19, folder 14–15, WCAP.

38. Allee to Salt, 8 March 1927, box 1, folder 13, Zoology Department Records, Joseph Regenstein Library, University of Chicago (hereafter ZDR). On Salt's corn borer study, see his *Report on Sugar-Cane Borers*. On Salt's work at Farnham House Laboratory, see W. R. Thompson, "Biological Control," 325–26; Kingsland, *Modeling Nature,* 132.

39. For biographical information on Emerson, see E. O. Wilson and Michener, "Alfred Edwards Emerson"; and T. Park, "Alfred Edwards Emerson."

40. Allee to Lillie, 9 June 1932, box 19, folders 14–15, WCAP.

41. Lillie to Mason, 5 June 1931, and Lillie to Mason, 20 February 1932, RG1.1 216D, box 8, folder 104, RFA. For information on Rockefeller funding to the Division of Biological Sciences at Chicago, see budget expenditure statements in RG1.1 216D, box 8, folders 103–9, RFA.

42. Lillie to Hutchins, 10 October 1933, wooden box, FRLP-MBL. For a history of the Rockefeller Foundation and its reorganization, see Fosdick, *Story of the Rockefeller Foundation;* and Kohler, "A New Policy."

43. Weaver, "The Program in Vital Processes," 24 October 1934,

IIA.106, FRLP-MBL. WW Diary, 18–19 January 1934, RG1.1 216D, box 8, folder 106, RFA. Warren Weaver's program is discussed in Kohler, "Warren Weaver."

44. The shift in funding is discussed in a number of Rockefeller memos. See RBF memo, 16 February 1937, RG1.1 216D, box 8, folder 108, RFA; Weaver to Lillie, 6 October 1933; Weaver memo, 8–11 September, 1933; 6 November 1933, RG1.1 216D, box 8, folder 105, RFA. Allee resented the withdrawal of his Rockefeller grant, and the situation was exacerbated when he tried to get support from the NRC Committee for Research in Problems of Sex. For his initial response, see Allee to Lillie, 10 December 1934, box 19, folder 14–15, WCAP.

45. Yerkes to Allee, 23 March 1935, box 9, folder 2, WCAP. The incident appears in a number of letters and memos. See Allee to Yerkes, 11 March 1935; memo concerning NRC grant for 1935–36; Yerkes to Allee, 6 March 1936, box 9, folder 2, WCAP. See also Yerkes to Weaver, 11 February 1936; Allee to Yerkes, 8 February 1936; Weaver to Yerkes, 28 February 1936, RG1.1 200, box 39, folder 440, RFA.

46. F. R. Lillie, "Proposal for an Institute of Genetic Biology in the University of Chicago," 29 January 1934, RG1.1 216D, box 8, folder 106, RFA.

47. Weaver to Lillie, 29 March 1934, RG1.1 216D, box 8, folder 106, RFA. Lillie to Flexner, 10 November 1934, IIA.106, FRLP-MBL.

48. On Weaver's shift in focus, see Kohler, "Warren Weaver."

49. See memo regarding conversation between Lillie and Allee, 13 March 1934, box 19, folder 14–15, WCAP.

50. Park's recollections as a graduate student and faculty member confirm the influence embryological research played in the department. Park recalled that Allee and his students "were treated as second-class citizens" in the department during the early thirties, but when Park returned in the late thirties, many people in embryology had retired, opening the door for ecology (author's interview with Thomas Park, 23 July 1985).

Chapter Six

1. Provine has argued that population geneticists provided the substantive framework for the modern synthesis. See his *Origins of Theoretical Population Genetics* and "Role of Mathematical Population Genetics." Mayr has stressed the importance of the naturalist tradition. See his "Where Are We?", 307–28. For anthologies on the historical origins of the synthesis, see Mayr and Provine, eds., *Evolutionary Synthesis;* and Grene, ed., *Dimensions of Darwinism.*

2. Mayr presents these five figures as the "architects of the Synthesis" in *Growth of Biological Thought,* 567.

3. See Emerson to Schmidt, 10 January 1919, box 6, folder 4, KPSP.

4. For an account of the early researches and founding of the Tropical Research Station, see Beebe, Hartley, and Howes, *Tropical Wild Life in British Guiana.* Beebe's opposition to the use of the station as a collecting site is dis-

cussed in Emerson to Schmidt, 4 March 1919, box 6, folder 4, KPSP. On Beebe's standing among professional biologists, see Emerson to Schmidt, 19 September 1920, box 6, folder 4, KPSP.

5. Emerson to Schmidt, 19 September 1920, box 6, folder 4, KPSP. For an autobiographical account of Emerson's early interest in termites, see Emerson to Collias, 23 April 1974, C file, Alfred Edwards Emerson Papers, Department of Entomology, American Museum of Natural History, New York (hereafter AEE-AMNH). Emerson also mentions his contact with Wheeler at Kartabo in M. A. Evans and H. E. Evans, *William Morton Wheeler, Biologist,* 276–77.

6. On the hiring of Emerson at Pittsburgh, see Emerson to Schmidt, 9 January 1921; Emerson to Schmidt, 5 May 1921; and Emerson to Schmidt, 7 June 1921, box 6, folder 4, KPSP.

7. Emerson to Schmidt, 22 June 1926, box 6, folder 4, KPSP.

8. Emerson, "Development of a Soldier of *Nasutitermes.*" For a review article on the theories of caste determination in the 1920s, see Snyder, "Biology of the Termite Castes."

9. Hare, "Caste Determination," 281. On Thompson's studies, see C. B. Thompson, "Origin of the Castes of the Common Termite"; and idem, "Development of the Castes."

10. Castle, "Experimental Determination of Caste Differentiation in Termites," 314; "Experimental Investigation of Caste Differentiation." For a brief historical overview of Castle's work, see E. M. Miller, "Caste Differentiation in the Lower Termites."

11. E. M. Miller, "Problem of Castes and Caste Differentiation."

12. For Wheeler's final statement concerning caste differentiation in ants, see his *Mosaics and Other Anomalies.* Gregg's dissertation research was published as "Origin of Castes in Ants."

13. Gregg, "Origin of Castes in Ants," 306–7.

14. Emerson, "Social Coordination," 186.

15. Wright, "Review of *The Origin and Development of the Nervous System.*" For a discussion of Wright's work in physiological genetics, see Provine, *Sewall Wright,* 160–206.

16. Wright, "Genetics of Abnormal Growth in the Guinea Pig," 142.

17. Ibid., 138, 139, 142.

18. Even with Hamilton's theory of kin selection, group selection still appears to be the most probable explanation of termite sociality since termites do not possess a haplo-diploid system of heredity.

19. Early papers (before 1950) that emphasized the value of the superorganism concept include Emerson, "Social Coordination"; idem, "Basic Comparisons of Human and Insect Societies"; idem, "Why Termites?"

20. Emerson, "Populations of Social Insects," 295.

21. The number of papers in which Emerson cited Wright for support of population-level selection is substantial. See, e.g., Emerson, "Termite Nests"; idem., "Social Coordination"; idem., "Ecology, Evolution, and Society"; W. C. Allee, et al., *Principles of Animal Ecology,* 683–95.

22. See Wright, "Evolution in Mendelian Populations"; and idem, "Role of Mutation." For a detailed discussion of the development of Wright's shifting balance theory, see Provine, *Sewall Wright,* 277–326.

23. Ronald A. Fisher, *Genetical Theory of Natural Selection.* On the differences between Fisher's and Wright's views, see Provine, *Sewall Wright,* 232–76.

24. Theodosius Dobzhansky, *Genetics and the Origin of Species,* 133, 15. Beatty has discussed the extent to which Dobzhansky relied on population adaptation arguments. See Beatty, "Dobzhansky and the Significance of Genetic Load."

25. Cain and Sheppard, "Theory of Adaptive Polymorphism."

26. Emerson, "Termitophile Distribution."

27. Ibid., 392.

28. Emerson, "The Origin of Species," 153. See also idem, "Taxonomic Categories and Population Genetics"; Allee et al., *Principles of Animal Ecology,* 605–30; Emerson, "Biological Species."

29. The founding of the Society for the Study of Speciation was announced in *American Naturalist* 75 (1941): 86–89. For Emerson's reviews of key works in the synthesis, see his "Origin of Species"; idem, "Systematics and Speciation"; idem, "Taxonomy and Ecology."

30. Ernst Mayr, "Speciation Phenomena in Birds."

31. Emerson to Mayr, 30 April 1941, folder E, 1940–43, Ernst Mayr Correspondence, Harvard University Archives, Cambridge, Mass.

32. For a historical discussion of other ecologists' views of selection during this period, see Kimler, "Advantage, Adaptiveness, and Evolutionary Ecology"; and Collins, "*Evolutionary Ecology* and the Use of Natural Selection."

33. Williams, *Adaptation and Natural Selection,* 130.

34. David Lack's *Natural Regulation of Animal Numbers* is a prime example of an ecological work in the 1950s couched in an individual selectionist mode of argument, although Lack's views were still by no means those of the majority.

35. Emerson, "Termite Nests."

36. Emerson, "Ethospecies," 10. See also idem, "Termite Nests."

37. Emerson, "Ethospecies." Emerson's student, Robert S. Schmidt, in "The Evolution of Nest-Building Behavior," followed Emerson's lead in using nest patterns to arrive at species evolution in the genus *Apicotermes.*

38. For a discussion of the emergence of mathematical population ecology, see Kingsland, *Modeling Nature.* The growing interest in population studies during this period is evidenced by the numerous symposia on population problems throughout the 1930s. See, e.g., "Symposium on Experimental Populations," *American Naturalist* 71 (1937); "Symposium on Insect Populations," *Ecological Monographs* 9 (1939); and "Symposium on Population Problems in Protozoa," *American Naturalist* 75 (1941). There were also a number of ecology texts, such as Chapman, *Animal Ecology,*" and Bodenheimer, *Problems in Animal Ecology,* that had a decided population focus. For an anecdotal history of Elton's Bureau of Animal Population, see Crowcroft, *Elton's Ecologists.*

39. See Shaw, "Effect of Biologically Conditioned Water upon Rate of Growth in Fishes and Amphibia"; W. C. Allee, Bowen, Welty, and Oesting, "Effect of Homotypic Conditioning of Water on the Growth of Fishes"; W. C. Allee, Oesting, and Hoskins, "Is Food the Effective Growth-Promoting Factor?" and G. Evans, "Relation between Vitamins and the Growth and Survival of Goldfishes."

40. The first evidence of tin contamination is discussed in W. C. Allee, Finkel, and Hoskins, "Growth of Goldfish in Homotypically Conditioned Water"; Finkel and W. C. Allee, "Effect of Traces of Tin." For evidence of later contamination, see W. C. Allee, P. Frank, and Berman, "Homotypic and Heterotypic Conditioning."

41. See, e.g., Robertson, *Chemical Basis of Growth and Senescence;* and Pearl, *Biology of Population Growth.* For a historical discussion of the logistic curve and the role of Robertson and Pearl, see Kingsland, *Modeling Nature,* 65–94.

42. See Robertson, "I. Multiplication of Isolated Infusoria," and "II. Influence of Mutual Contiguity"; idem, "Reproduction in Cell-Communities"; idem, "CLXII. Allelocatalytic Effect in Cultures of *Colpidium*"; idem, "Influence of Washing"; idem, "On Some Conditions Affecting the Viability of Cultures of Infusoria."

43. See, e.g., Cutler and Crump, "Rate of Reproduction in Artificial Cultures of *Colpidium colpoda,*" I, II, and III; idem, "Influence of Washing on the Reproductive Rate of *Colpidium colpoda,* II"; Greenleaf, "Influence of Volume of Culture Medium . . . on . . . Protozoa"; idem, "Influence of Volume of Culture Medium . . . on . . . Infusoria"; Jahn, "Studies on the Physiology of the Euglenoid Flagellates."

44. See, e.g., Petersen, "Relation of Density of Population"; Johnson, "Effects of Population Density"; Sweet, "Micropopulation of *Euglena gracilis* Klebs."

45. Johnson, "Effects of Population Density," 52.

46. Pearl, Miner, and Parker, "Experimental Studies on the Duration of Life"; Bodenheimer, *Problems in Animal Ecology,* 56.

47. Chapman, "Quantitative Analysis of Environmental Factors."

48. W. C. Allee, *Animal Aggregations,* 178–80.

49. T. Park, "Studies in Population Physiology," I and II.

50. Allee to Park, 5 January 1934, box 20, folders 17–19, WCAP. For additional correspondence pertaining to the dispute between Chapman and Park, see Park to Allee, 3 January 1934, 17 February 1934; Park to Allee, 8 October 1934; Allee to Park, 27 April 1935; and Park to Allee, 18 January 1937, box 20, folder 17–19, WCAP.

51. See Chapman, "Quantitative Analysis"; idem, *Animal Ecology.*

52. Park to Allee, 3 January 1934, box 20, folders 17–19, WCAP. See also Chapman, "Causes of Fluctuation"; idem, "Experimental Analysis."

53. Elton, *Animal Ecology.*

54. For some of the important works on this issue during the 1930s, see Bodenheimer, *Problems in Animal Ecology,* 81–112; Nicholson, "Balance of

Animal Populations"; H. S. Smith, "Role of Biotic Factors"; W. R. Thompson, "Biological Control"; Uvarov, "Insects and Climate." For secondary historical literature on various aspects of this issue, see Collins, "*Evolutionary Ecology* and the Use of Natural Selection"; Kimler, "Advantage, Adaptiveness, and Evolutionary Ecology"; idem, "Mimicry"; Kingsland, *Modeling Nature,* 116–23; and McIntosh, *Background of Ecology,* 189–90.

55. W. R. Thompson, "Biological Control," 328.

56. "Discussion at Staff Meeting held December 2, 1935," box 9, folder 6, WCAP. See also Allee to Wright, Election Day, 1936, box 24, folder 11, WCAP.

57. See Pearl, "On Biological Principles Affecting Populations." Pearl's approach is also discussed in Kingsland, *Modeling Nature,* 80–83.

58. W. C. Allee, *Social Life of Animals,* 37.

59. See esp. W. C. Allee, "Integration of Problems."

60. Ludwig and Boost, "Über das Wachstum von Protistenpopulationen"; Hutchinson, "A Note on the Theory of Competition"; W. C. Allee and Odum, "A Note on the Stable Point of Populations."

61. W. C. Allee, "Concerning the Origin of Sociality in Animals," 158.

62. W. C. Allee, "Integration of Problems," 485.

63. For Allee's citations of Wright's work, see, e.g., Allee, *Social Life of Animals,* 97–111; idem, "Concerning the Origin of Sociality," 156; idem, "Integration of Problems," 485; idem, "Human Conflict and Co-operation," 328–30. Kimler, in "Advantage, Adaptiveness, and Evolutionary Ecology," 225, has also pointed out the importance of Wright's work for the Chicago ecologists' views.

64. W. C. Allee, "Concerning the Origin of Sociality," 155.

65. T. Park, "Laboratory Population as a Test," 457. In a personal interview with the author (23 July 1985), Park told me that he felt the superorganism concept had little, if any, analytical power. The similarity of his views to Allee's and Emerson's, however, is particularly apparent when compared to other approaches. Lerner and Dempster, for example, analyzed the indeterminacy in the competition experiments of Park and associates by looking at within-species genetic variation, analyzing within-population influences rather than external factors that affect the population as a unit. See Lerner and Dempster, "Indeterminism in Interspecific Competition." Collins, in "*Evolutionary Ecology* and the Use of Natural Selection," 272–73, also discusses this experiment.

66. Allee to Lillie, 5 August 1938, box 19, folder 14–15, WCAP.

67. Allee was president of the ESA in 1929, Emerson was president in 1941, and Park was president in 1958.

68. W. C. Allee and T. Park, "Concerning Ecological Principles." For information regarding the "Ecology Group," see K. P. Schmidt, "Warder Clyde Allee," 16–18. This information also came from the author's interview with Thomas Park, 23 July 1985.

69. Ann Allee to Schmidt, 18 December 1955, box 1, folder 2, KPSP; K. P. Schmidt, "Warder Clyde Allee," 20.

70. Actual sales of the book have been difficult to determine. From royalty slips, it appears that 5,877 copies were sold through September 1954. See W. B. Saunders file, KPSP. W. C. Allee et al., *Principles of Animal Ecology*, 6.

71. Odum, *Fundamentals of Ecology*. For a historical analysis of the rise of ecosystems ecology and a comparison of *Principles of Animal Ecology* with *Fundamentals of Ecology*, see Kwa, "Mimicking Nature." See also idem, "Representations of Nature"; McIntosh, *Background of Ecology*, 193–241; and P. J. Taylor, "Technocratic Optimism."

72. Hanson to Allee, 24 January 1941, box 9, folder 3, WCAP. Much of the correspondence pertaining to the Committee on the Ecology of Animal Populations is available in box 2, folders 2–4; box 9, folder 3, WCAP; and RG1.1 216D, box 10, folders 130–38, RFA.

73. For notes of the first meeting, see "Proceedings of the Meeting of the Committee on the Ecology of Animal Populations," box 2, folder 2, WCAP.

74. "Proceedings of the Meeting of the Committee on the Ecology of Animal Populations," p. 2; Johnson to Park, 1 March 1941, box 2, folder 2, WCAP. The actual list of problems appears in the final proposal submitted to the Rockefeller Foundation. See, "Organization of Population Problems and Investigators," App. 1 in "The Status and Need for Further Research in the Ecology of Animal Populations," RG1.1 216D, box 10, folder 138, RFA, 13.

75. "The Status and Need For Further Research," RFA.

76. "The Status and Need for Further Research," RFA, 6–7; App. 8, "An Attempt at a Working Basis for the Delimitation of Population Ecology and General Ecology," RFA, 3.

77. Hanson to Allee, 15 May 1942, box 9, folder 3, WCAP.

78. Allee to Hanson, 25 May 1942, RG1.1 216D, box 10, folder 131, RFA; "Status and Need for Further Research," title page; Hanson to Allee, 15 September 1942, RFA.

79. For the renewed attempt to get funding for the committee after the war, see box 2, folder 6, WCAP.

80. Hutchinson and Wollack, "Studies on Connecticut Lake Sediments," 510. For an autobiographical account of Hutchinson's life, see his *Kindly Fruits of the Earth*.

81. Hutchinson, "Bio-Ecology," 268. Hutchinson and Wollack, "Studies on Connecticut Lake Sediments," 510.

82. Lindeman, "The Trophic-Dynamic Aspect of Ecology."

83. Hutchinson to Allee, 30 January 1947, box 2, folder 6, WCAP.

84. Gause, *Struggle for Existence;* idem., "Principles of Biocoenology"; Hutchinson, "Ecological Aspects of Succession"; idem, "Limnological Studies in Connecticut." For a historical analysis of Gause's work and its impact on Hutchinson, see Kingsland, *Modeling Nature,* 146–80. See also Jackson, "Interspecific Competition and Species' Distributions"; McIntosh, *Background of Ecology,* 178–92.

85. Clements and Shelford, *Bio-Ecology*. Keller, in "Demarcating Public from Private Values," has documented the increase in the proportion of ecological papers using competition in the title or key words after World War II, although she sees the rise beginning in the late 1960s.

86. Allee et al., *Principles of Animal Ecology,* 495–528.

87. On Park's role in the publication of Lindeman's paper, see R. E. Cook, "Raymond Lindeman." For a representative sample of Park's research papers on competition, see T. Park, "Laboratory Population"; idem, "Experimental Studies of Interspecies Competition"; idem, "Beetles." For a philosophical analysis of Park's experiments in the context of group selection, see Griesemer and Wade, "Laboratory Models." For the list of Park's RF funding support, see RG 1.1 216D, box 10, folder 130, RFA.

88. Griggs to Allee, 31 January 1947, box 2, folder 6, WCAP.

89. Hutchinson and Deevey, Jr., "Ecological Studies on Populations," 332.

90. Weaver Diary, 17 January 1950, RG 1.1 216D. box 10, folder 135, RFA.

91. For an account of Allee's RF funding at the University of Chicago, see RG 1.1 216D, box 10, folder 130, RFA. For his RF support at the University of Florida, see RG1.2 Series 200D, box 229, folder 2210, RFA.

92. Allee's social behavior research is the subject of chap. 8. For a discussion of his influence on the study of animal behavior, see Banks, "Allee and the Chicago School of Animal Behavior"; and Collias, "Role of American Zoologists."

Chapter Seven

1. Hutchins, "What Shall We Defend?" 549.

2. Purcell, *Crisis of Democratic Theory.*

3. Adler, "Chicago School."

4. Kuznick, *Beyond the Laboratory.*

5. Friley, "Science and Human Values," 71.

6. Fosdick, "Science and the Moral Order," 306.

7. Conklin, "Science and Ethics," 199. For a discussion of events leading to this Magna Charta and the significance of the AAAS meetings at Indianapolis, see Kuznick, *Beyond the Laboratory,* 64–82.

8. See, e.g., Dewey, "Antinaturalism in Extremis"; and Hook, "Naturalism and Democracy." Hutchins's attack on Dewey and Dewey's equating of the scientific method with democratic government is discussed in Purcell, *Crisis of Democratic Theory,* 150–52, 200–207.

9. Conklin, *Man, Real and Ideal,* 171, 196.

10. Conklin, "Does Science Afford a Basis for Ethics?"; S. J. Holmes, "Darwinian Ethics"; Herrick, "A Neurologist Makes Up His Mind." For other articles advocating a biologically based ethics during this period, see Montagu, "The Socio-biology of Man"; Brody, "Science and Social Wisdom"; Brown, "Ecological Concepts and Human Welfare"; Child, "Social Integration"; J. S. Huxley, "Biologist Looks at Man"; idem, *Touchstone for Ethics;* Sinnott, "Biological Basis of Democracy."

11. "Section on Historical and Philological Sciences," *Science* 93 (1941): 136; Leake, "Ethicogenesis," 253.

12. Waddington's article and the series of commentaries that appeared in *Nature* are reprinted in Waddington, *Science and Ethics,* 18.

13. Redfield, ed., *Levels of Integration.* The plans for this symposium are discussed in Redfield's introduction to the volume. Information also came from the author's interview with Thomas Park, 23 July 1985.

14. Haraway, in "Biological Enterprise," introduces the term "organic functionalism" to characterize the naturalistic assumptions and metaphors inherent in the life sciences during the interwar years. For a discussion of organic functionalism in other biological disciplines, see Cross and Albury, "Walter B. Cannon, L. J. Henderson, and the Organic Analogy"; Haraway, *Crystals, Fabrics, and Fields;* idem, "Signs of Dominance"; Russett, *Concept of Equilibrium.*

15. T. Park, "Integration in Infra-social Insect Populations"; Emerson, "Populations of Social Insects."

16. Emerson, "Ecology, Evolution and Society," 101.

17. Emerson to Allee, 5 February 1954, box 16, folders 9–10, WCAP.

18. See, e.g., Child, *Physiological Foundations of Behavior;* idem, "Social Integration"; and idem, *Patterns and Problems of Development.* For secondary accounts of Child's physiological theory of dominance, see Haraway, *Crystals, Fabrics, and Fields,* 53–54; and idem, "Signs of Dominance," 144–45.

19. Emerson, "Social Coordination," 187.

20. Emerson, "Populations of Social Insects," 294.

21. Cross and Albury, "Walter B. Cannon, L. J. Henderson, and the Organic Analogy," 186.

22. Cannon, "Physiological Regulation of Normal States."

23. Cannon, *Wisdom of the Body,* 300.

24. Cross and Albury, "Walter B. Cannon, L. J. Henderson, and the Organic Analogy," 168.

25. The literature of the period touching on this question is immense. See, e.g., the 1935 symposium of the American Academy of Political and Social Science entitled "Socialism, Fascism and Democracy," *Annals of the American Academy* 180 (1935); and the essays in Anshen, *Freedom.*

26. The impact of the Popular Front and the rise of fascism on American intellectuals is discussed in Pells, *Radical Visions,* 292–329.

27. Emerson published only two papers that experimentally analyzed the ecological functions of the termite nest. See Emerson, "Termite Nests"; idem, "Regenerative Behavior." The two doctoral dissertations written on this subject under Emerson's direction were Talbot, "Distribution of ant species in the Chicago region"; and Scherba, "Microclimatic modification."

28. In the author's interviews with students of Emerson (David Kistner on 7 January 1987 and Gerald Scherba on 9 January 1987) both mentioned that while Emerson almost never spoke negatively about other colleagues, he sharply criticized Hutchins and Adler for their antiscience attitudes. This is also apparent in a letter Emerson wrote to Ralph Burhoe commenting on Adler: "I know Mortimer Adler personally. . . . I have no confidence whatever in his

antiscientific attitudes. I think he is the last person to resolve any of these problems [Emerson was speaking of a naturalistic basis of ethics] by means of a scientific approach." Emerson to Burhoe, 16 December 1963, Alfred Edwards Emerson Papers, Joseph Regenstein Library, University of Chicago (hereafter AEE-UC).

29. Emerson, "Biological Sociology," 146. See also idem, "Basic Comparisons of Human and Insect Societies"; idem, "Ecology, Evolution and Society"; and idem, "Biological Basis of Social Cooperation."

30. Emerson, "Basic Comparisons of Human and Insect Societies," 168–69.

31. Emerson, "Biological Basis of Social Cooperation," 12.

32. See W. C. Allee, "Cooperation and Conflict as Modes of Social Integration"; Emerson, "Cooperation and Conflict in the Balance of Nature"; Gerard, "Cooperation and Conflict as Modes of Social Integration in Levels of Organization," typescript, Allee Reprint Collection, Earlham College, Richmond, Ind.

33. Emerson, "Biological Basis of Social Cooperation," 15.

34. Ibid., 15.

35. Ibid., 16.

36. Ibid., 17–18.

37. Emerson, "Biological Foundations of Ethics and Social Progress," 288.

38. Emerson, "Why Termites?" 345.

39. "Transcript of Alfred E. Emerson's Seminar," Winter 1967, AEE-UC, 51.

40. For biographical information on Gerard, see Seymour S. Kety, "Ralph Waldo Gerard."

41. These included Gerard, "Organism, Society and Science"; idem, "Organic Freedom"; idem, "Biological Basis for Ethics"; idem, "Science and the Public"; idem, "Biologist's View of Society."

42. Kuznick, *Beyond the Laboratory,* 228. For a discussion of Gerard's activities in the AASW, see Kuznick, *Beyond the Laboratory,* 166–70, 231, 237, 240, 246.

43. Gerard, "Organism, Society and Science," 408.

44. Gerard, "Higher Levels of Integration," 79.

45. Gerard, "Organism, Society and Science," 532.

46. Ibid., 411.

47. Gerard, "Higher Levels of Integration," 82. See also idem, "Organic Freedom," 424–25; idem, "Organism, Society and Science," 534–35.

48. Graebner, *Engineering of Consent,* 128.

49. Simpson, "Role of the Individual in Evolution," 19.

50. Emerson to Burhoe, 16 December 1963, AEE-UC.

51. Simpson, "Role of the Individual," 16.

52. Ibid., 19.

53. Simpson, *Meaning of Evolution,* 311. Greene, in *Science, Ideology, and World View,* 168–79 provides a comprehensive discussion of Simpson's

evolutionary ethics, although Greene does not really analyze the importance of the individual in Simpson's writings.

54. Simpson, "Role of the Individual," 20.

55. For a historical analysis of Novikoff's career in the context of McCarthyism, see D. R. Holmes, *Stalking the Academic Communist.*

56. Novikoff, "Concept of Integrative Levels and Biology," 213.

57. Schneirla, "Problems in the Biopsychology of Social Organization." For biographical information on Schneirla, see Ethel Tobach and Lester R. Aronson, "Biographical Note," xi–xviii; and Roback, "Theodore C. Schneirla." Stephen Weldon, in "Integrating the Ant," briefly discusses Schneirla's support of left-wing causes during the 1930s and 1940s. I thank Stephen for calling my attention to the following House reports in which Schneirla's name appears. See Committee on Un-American Activities, *Communist Political Subversion, Pt. 2, Appendix to Hearings,* 84th Cong., 2nd sess., 1957, pp. 7185–95, 7292; *Testimony of Bishop G. Bromley Oxnam,* 83rd Cong., 1st sess., 21 July, 1953, pp. 3638–40, 3657–88; *Civil Rights Congress,* 80th Cong., 1st sess., 2 September 1947, House Report no. 1115, pp. 26–28; and *Report on the Communist "Peace" Offensive: A Campaign to Disarm and Defeat the United States,* 82nd Cong., 1st sess., 1 April 1951, House Report no., 378, p. 150.

58. Novikoff, "Concept of Integrative Levels," 214.

59. Gerard and Emerson, "Extrapolation from the Biological to the Social," 584. See also the exchange of correspondence between Emerson and Novikoff in Emerson to Novikoff, 15 April 1945; Novikoff to Emerson, 19 March 1945, AEE-UC.

60. Novikoff, "Continuity and Discontinuity."

61. Novikoff, "Concept of Integrative Levels," 214.

62. Emerson was quite explicit about the importance of social control in his political outlook. Writing to his former student, Kumar Krishna, in 1969, Emerson commented, "I noticed today that Indira Ghandi's candidate for President won. I hope this is good. She is considered leftish, but I doubt she is a real communist in either sympathy or attitude. She is a socialist. I am not, but at one time I thought I was until I found out more about their achievements. Their ideals are good, but their implementation is seldom as good as the results of competative [*sic*] capitalism (under some strong social control)." Emerson to Krishna, 21 August 1969, AEE-AMNH.

Chapter Eight

1. Haraway, "Animal Sociology," 33.

2. The chairmen who were University of Chicago graduates were J. P. Scott, A. M. Guhl, W. C. Young, N. E. Collias, and E. B. Hale. See Collias, "Role of American Zoologists."

3. These included N. E. Collias, A. E. Emerson, E. B. Hale, E. Hess, J. P. Scott, and W. C. Young.

4. On the development of continental ethology, see Burkhardt, "Devel-

opment of an Evolutionary Ethology"; idem, "On the Emergence of Ethology as a Scientific Discipline"; Durant, "Innate Character in Animals and Man."

5. T. Schjelderup-Ebbe, "Beitrage zur Sozial-psychologie des Haushuhns"; and idem, "Social Behavior of Birds."

6. For a listing of animal behavior research funded by the National Research Council Committee, see Aberle and Corner, *Twenty-Five Years of Sex Research*. A review of the literature on dominance during this period is available in Meredith P. Crawford, "Social Psychology of the Vertebrates"; and Collias, "Aggressive Behavior among Vertebrate Animals." For a historical analysis of dominance research in primatology, see Haraway, *Primate Visions;* idem, "Signs of Dominance."

7. W. C. Allee, "Animal Aggregations: A Request for Information"; Masure and W. C. Allee, "Social Order in Flocks of the Common Chicken."

8. Masure and W. C. Allee, "Flock Organization of the Shell Parrakeet," 397.

9. For a historical account of endocrinological research on behavior during this period, see Beach, "Historical Origins of Modern Research on Hormones and Behavior."

10. Yerkes to Allee, 23 March 1935. The incident appears in a number of letters and memos. See Allee to Yerkes, 11 March 1935; Memo concerning NRC grant for 1935–36; Yerkes to Allee, 6 March 1936, all box 9, folder 2, WCAP. See also Yerkes to Weaver, 11 February 1936; Allee to Yerkes, 8 February 1936; Weaver to Yerkes, 28 February 1936, RG1.1 200, box 39, folder 440, RFA. For Allee's attempts to get research funds from other agencies, see box 6, folder 1, WCAP; Allee to Merriam, 11 January 1937; Allee to Redfield, 16 April 1936, box 9, folder 2, WCAP.

11. See, e.g., Noble, Wurm, and A. Schmidt, "Social Behavior of the Black-Crowned Night Heron"; Noble, Kumpf, and Billings, "Induction of Brooding Behavior in the Jewel Fish"; Noble and Wurm, "Effect of Testosterone Proprionate on the Black-Crowned Night Heron"; Noble and Greenberg, "Induction of Female Behavior in the Male *Anola carlinensis*"; Noble and Zitrin, "Induction of Mating Behavior in Male and Female Chicks."

12. W. C. Allee, Collias, and Lutherman, "Modification of the Social Order in Flocks of Hens." See also W. C. Allee and Collias, "Effect of Estradiol"; W. C. Allee, Collias, and Beeman, "Effect of Thyroxin."

13. Ginsburg and W. C. Allee, "Some Effects of Conditioning," 505.

14. See Allee to Buchsbaum, 23 September 1943, box 14, folder 4, WCAP.

15. Beeman and W. C. Allee, "Some Effects of Thiamin," 220.

16. Richard Francis has also argued that the distinction between the sociological perspective of Chicago animal behavior studies and the atomistic accounts of social behavior that have become characteristic of ethology and sociobiology is significant. See Francis, "On the Relationship between Aggression and Social Dominance."

17. Guhl and W. C. Allee, "Some Measurable Effects," 320.

18. Ibid., 343.

19. Greenberg, "Some Relations between Territory, Social Hierarchy, and Leadership."

20. Guhl and Allee, "Some Measurable Effects," 346.

21. Tinbergen, *Social Behavior in Animals,* 71.

22. These included Emlen, Jr. and F. W. Lorenz, "Pairing Responses of Free-Living Quail"; Howard and Emlen, Jr., "Intercovey Social Relationships"; Odum, "Annual Cycle of the Black-Capped Chickadee, I"; and J. W. Scott, "Mating Behavior of the Sage Grouse."

23. For a review of the territoriality literature, see Nice, "The Role of Territory in Bird Life."

24. Nice, review of *The Social Life of Animals,* by W. C. Allee, *Bird-Banding* 9 (1938): 218.

25. See, e.g., Shoemaker, "Social Hierarchy in Flocks of the Canary"; Ritchey, "Dominance-Subordination and Territorial Relationships in the Common Pigeon"; and Greenberg, "Some Relations."

26. Allee to Tinbergen, 23 December 1953, box 23, folder 1, WCAP.

27. For a discussion of these institutes, see Chatfield, *For Peace and Justice,* 137–38; DeBenedetti, *Peace Reform in American History,* 121.

28. See "Some Biological Aspects of International Relations" (1935), Allee Correspondence 1905–38, EBC; box 26, folder 10, WCAP.

29. M. H. Allee, *Camp at Westlands,* 77.

30. Commencement Address, Earlham College, 1934, box 5, folder 9, WCAP, 12–13.

31. W. C. Allee, *Social Life of Animals,* 188.

32. Ibid., 211.

33. Ibid., 213.

34. Ibid., 214.

35. Earlham College, Commencement Address, 1934, WCAP, 9.

36. M. H. Allee to Molly Barth, 25 February 1942 from a letter Molly Barth sent to the author, 10 March 1990.

37. For an account of Pelkwyk's associations with Margaret Nice, see Nice, *Research is a Passion with Me,* 220–32.

38. Allee to Pelkwyk-Donath, 5 November 1945, box 21, folder 2, WCAP.

39. Allee to Taylor, 25 October 1940; see also Allee to Taylor, 5 February 1940; Taylor to Allee, 8 February 1940; 10 October 1940, box 23, folder 4, WCAP.

40. See, Nelson to Allee, 3 February 1948, box 20, folder 11, WCAP. Allee to Nelson, 4 May 1948, Allee Correspondence, 1942–55, EBC; Williams to Allee, 5 January 1948, box 24, folder 4, WCAP.

41. Allee to Taylor, 18 August 1942; see also Taylor to Allee, 9 June 1942, 11 August 1942, box 23, folder 4, WCAP.

42. See Elisabeth M. Borgese, "Animal Kingdom and the World Republic"; Borgese to Allee, 7 January 1950, box 13, folder 4, WCAP.

43. Allee to Hutchins, 14 July 1945, box 19, folder 12, UCPP, 1940–46.

44. W. C. Allee, "Where Angels Fear to Tread." This article was reprinted in *An Editor's Notebook* 1 (1943): 5–12; *Current Religious Thought* 3 (1943): 19–29; *Nature* 152 (1943): 309–12; *Main Currents in Modern Thoughts* (1944): and *Sociometry* 8 (1945): 21–29.

45. Allee, "Where Angels Fear to Tread," 519.

46. Ibid., 521.

47. Ibid., 523.

48. Ibid., 524.

49. Coker, "What Are the Fittest?" 488. For similar statements, see Montagu, "Nature of War and the Myth of Nature"; Brody, "Science and Social Wisdom"; Brown, "Ecological Concepts"; Keith, *Evolution and Ethics;* S. J. Holmes, *Life and Morals;* J. S. Huxley, "War as a Biological Phenomenon," 60–68.

50. T. H. Huxley, "Evolution and Ethics." Although there has been historical debate about the meaning of Huxley's essay, it is quite clear that Allee interpreted Huxley's remarks as a separation between biology and ethics. For insight into this debate, see Helfand, "T. H. Huxley's 'Evolution and Ethics.'"

51. Allee, "Where Angels Fear to Tread," 524.

52. Ibid., 525.

53. W. C. Allee, "Biology and International Relations," 816.

54. Darling, *Herd of Red Deer.*

55. W. C. Allee, M. N. Allee, Ritchey, and Castles, "Leadership in a Flock of White Pekin Ducks."

56. Earlham College, Commencement Address, 1934, WCAP, 10.

57. W. C. Allee, "Human Conflict and Co-operation," 359.

58. W. C. Allee, "Dominance and Hierarchy in Societies of Vertebrates," 177.

59. For a brief synopsis of this research, see J. P. Scott and Fuller, "Research on Genetics and Social Behavior." For a historical analysis of the dog behavior studies in the context of scientific patronage and behavioral genetics, see Paul "Rockefeller Foundation."

60. Gregg to Little, 3 January 1944, RG1.2 200A, box 133, folder 1189, RFA.

61. For an autobiographical account of Scott's career, see J. P. Scott, "Investigative Behavior."

62. J. P. Scott, "Embryology of the Guinea Pig."

63. For a discussion of the research in developmental biology at Chicago and its opposition to genetic determinist theories, see Scott F. Gilbert, "Cellular Politics"; and Sapp, *Beyond the Gene.*

64. J. P. Scott, "Genetic Differences in the Social Behavior of Inbred Strains of Mice"; idem, "Experimental Test."

65. J. P. Scott, "Social Behavior."

66. Ibid., 19.

67. Ibid., 26.

68. Stewart and Scott, "Lack of Correlation between Dominance Relationships."

69. The studies Scott cited were Anderson, "Domination and Integration"; idem, "Experimental Study of Dominative and Integrative Behavior"; idem, "Domination and Social Integration"; idem, "Examination of the Concepts of Domination and Integration"; idem, "Studies in Dominative and Socially Integrative Behavior." See also Lewin, Lippitt, and White, "Patterns of Aggressive Behavior"; Lippitt, "Field Theory and Experiment in Social Psychology"; idem, "Experimental Study"; Pigors, *Leadership or Domination.* Some of these studies are discussed by Graebner, *Engineering of Consent,* within the context of democratic social engineering.

70. J. P. Scott, "Investigative Behavior, 405.

71. J. P. Scott, "Social Behavior, Organization, and Leadership," 1.

72. For historical background on this conference, see J. P. Scott, "Organization of Comparative Psychology." See also Scott to Gregg, 6 February 1946; Scott to Morison, 26 April 1946; Scott to Gregg, 2 July 1946; Scott to Gregg, 30 August 1946; RSM Interview, 10 September 1946; RG1.2 200A, box 134, folder 1190, RFA.

73. Dollard et al., *Frustration and Aggression,* 1.

74. J. P. Scott, "Dominance and the Frustration-Aggression Hypothesis," 37.

75. Ibid., 38.

76. J. P. Scott, *Aggression,* 80–81.

77. Ibid., 62.

78. G. Lerner, *Creation of Patriarchy,* 213.

79. The influence of the biological sciences in the naturalization of gender roles has received extensive treatment by feminist scholars. See, e.g., Bleier, *Science and Gender;* Birke, *Women, Feminism and Biology;* Fausto-Sterling, *Myths of Gender.*

80. Haraway, *Primate Visions.* See also idem, "Signs of Dominance"; idem, "Animal Sociology"; idem, "Biological Enterprise"; idem, "Teddy Bear Patriarchy."

81. Carpenter, "Field Study," 94–95, 99.

82. Haraway, "Animal Sociology," 33. See also idem, "Signs of Dominance."

83. J. P. Scott, *Aggression,* 80–88.

84. Biographical information on Noble is available in the *National Cyclopaedia of American Biography,* s.v. "Noble, Gladwyn Kingsley"; and in *Copeia* (1940):274–75; see also "Gladwyn Kingsley Noble," Biography II folder, Gladwyn Kingsley Noble Papers, Department of Herpetology, American Museum of Natural History, New York (hereafter GKNP).

85. On Gregory, see Ronald Rainger, "Vertebrate Paleontology as Biology."

86. Osborn to Noble, 17 March 1928; see also Noble to Osborn, 20 March 1928, AMNH Presidency, Henry F. Osborn I folder; and Sherwood to Noble, 18 May 1928, Department History folder, GKNP. On Morgan's move to the California Institute of Technology, see Allen, *Thomas Hunt Morgan,* 334–47.

87. Noble to Sherwood, 22 May 1928, Biography I folder; and Sherwood to Noble, 1 June 1928, Department: Budgets folder, GKNP.

88. For a brief description of the laboratory, see Noble to Burden, 15 October 1934, AMNH Departments: Experimental Biology folder, GKNP. On the political and social context of the African hall and the American Museum in general, see Haraway, "Teddy Bear Patriarchy."

89. See Sherwood to Noble, 13 January 1933; "Draft of Budget for 1934," 13 November 1933, Department: Budgets folder; and Faunce to Noble, 9 April 1935, AMNH Departments: Experimental Biology folder, GKNP.

90. An account of WPA help in the department and their respective duties can be found in Department: Personnel (WPA) folder, GKNP.

91. Noble to Burden, 15 October 1934, AMNH Departments: Experimental Biology folder, GKNP.

92. Noble to Burden, 18 June 1931, Burden, W. Douglas folder, GKNP.

93. Burden to Noble, n.d., Burden, Douglas W. folder, GKNP.

94. Noble, "Recent Advances in Our Knowledge of Sex," *New York World-Telegram,* June 1931.

95. The overall scope of Noble's research is evident from foundation progress reports. See esp. Foundations & Institutes: National Research Council Folder, and Foundations & Institutes: Josiah Macy, Jr. folder, GKNP. For representative publications, see Noble, *Biology of the Amphibia;* Noble and Curtis, "Social Behavior of the Jewel Fish"; Noble, Wurm, and A. Schmidt, "Social Behavior of the Black-Crowned Night Heron"; Noble, "Experimental Animal." For a historical account of the laboratory of experimental biology and Noble's behavioral research, see Mitman and Burkhardt, "Struggling for Identity."

96. Noble, "Experimental Animal," 118. See also idem, "Role of Dominance in the Social Life of Birds."

97. See Noble, Wurm, and Schmidt, "Social Behavior of the Black-Crowned Night Heron."

98. For a comparison with other degree-granting institutions during this period, see Rossiter, *Women Scientists in America.*

99. Harold W. Flitcraft, *History of the 57th Street Meeting of Friends,* 32. On housing discrimination in Chicago during the 1940s, see Hirsch, *Making the Second Ghetto.*

100. Allee to Waring, 7 June 1945, box 26, folder 10, WCAP.

101. Winslow, "Marjorie Hill Allee."

102. M. H. Allee, *The House,* 180.

103. For departmental discussions regarding the instructor position, see box 12, folder 16, WCAP. On Just and Young, see Manning, *Black Apollo of Science.*

104. M. H. Allee, *Great Tradition,* 111.

105. John Beatty, "Ecology and Evolutionary Biology," 262.

Chapter Nine

1. UNESCO, *Race Concept,* 103. On the formulation of the Unesco statements on race, see Haraway, *Primate Visions,* 197–203.

2. On the hardening of the synthesis, see Gould, "Hardening of the Modern Synthesis."

3. Emerson to Krishna, 21 August 1969, AEE-AMNH.

4. Wynne-Edwards, *Animal Dispersion;* Williams, *Adaptation and Natural Selection.* For historical reviews of the group selection controversy, see Brandon and Burian, eds., *Genes, Organisms, Populations,* 3–7; Gould, *Panda's Thumb,* 85–94; Sober, *Nature of Selection,* 215–26; Wade, "Critical Review"; D. S. Wilson, "Group Selection Controversy"; E. O. Wilson, "Group Selection."

5. Keller, "Demarcating Public from Private Values," 196.

6. "Human Cultural Evolution and Its Relation to Organic Evolution of Insect Societies," 1961, AEE-AMNH, 3–4. This manuscript was published in 1965 under the same title.

7. Keller, "Demarcating," 195.

8. Purcell, *Crisis of Democratic Theory,* 238.

9. Riesman, *Individualism Reconsidered,* 37.

10. Pells, *Liberal Mind in a Conservative Age,* 246.

11. Riesman, *Individualism Reconsidered,* 37–38.

12. Finney, *Body Snatchers.*

13. Riesman, *Individualism Reconsidered,* 117.

14. Ibid., 164, italics mine.

15. Ibid., 18.

16. T. Park, "Beetles," 1369. Park's first paper in his series on studies in population physiology was titled "Studies in Population Physiology." In 1948, he initiated a new series under the heading of interspecies competition. See T. Park, "Experimental Studies." For a philosophical analysis of Park's *Tribolium* system in the context of the units-of-selection controversy, see Griesemer and Wade, "Laboratory Models."

17. Simberloff, "Succession of Paradigms in Ecology," 78–82.

18. Dahl, *Preface to Democratic Theory,* 132. On Dahl and democratic theory in the cold war period, see Purcell, *Crisis of Democratic Theory,* 235–66.

19. Lack, *Natural Regulation of Animal Numbers.* On Lack, see Kingsland, *Modeling Nature,* 162–77; Collins, "*Evolutionary Ecology* and the Use of Natural Selection."

20. Robert H. MacArthur, "Fluctuations of Animal Populations." See also Hutchinson, "Homage to Santa Rosalia." For May's critique, see his *Stability and Complexity in Model Ecosystems.*

21. D. E. Davis, "Inquiry into the Phylogeny of Gangs," 319–20.

22. For an interesting perspective on the history of aggression studies, see Durant, "Beast in Man."

23. Ardrey, *Territorial Imperative,* 37.

24. Lorenz, *On Aggression,* 279, 281.

25. Many of the critical responses to Lorenz's *On Aggression* and Ardrey's *Territorial Imperative* can be found in Montagu, ed., *Man and Aggression.*

26. Lorenz, *On Aggression,* 276.

27. E. O. Wilson, *Sociobiology.* See also idem, *On Human Nature.*

28. On the transformation from organicist to cybernetic metaphors in the post-World War II era, see Haraway, *Primate Visions;* idem, "Signs of Dominance"; idem, "High Cost of Information"; P. J. Taylor, "Technocratic Optimism."

29. On Odum, ecosystem ecology, and cybernetics, see Kwa, "Mimicking Nature" 41–90. See also idem, "Representations of Nature."

30. Botkin, *Discordant Harmonies,* 190.

31. On the image of biologist as engineer, see Pauly, *Controlling Life.*

32. Worster, *Nature's Economy,* 347. For another recent appeal to cooperationist metaphors, see Augros and Stanciu, *New Biology.*

33. This essentialist genre of ecofeminism is in direct opposition to the more commonly held position within feminism that the gender categories masculine and feminine are themselves social constructs. For a critical analysis of this particular strand of ecofeminist literature, see Birke, *Women, Feminism, and Biology,* 115–25.

34. Hecht and Cockburn, *Fate of the Forest.*

Bibliography

Manuscript Sources

AEE-UC	Alfred Edwards Emerson Papers. Joseph Regenstein Library, University of Chicago, Chicago, Illinois.
AEE-AMNH	Alfred Edwards Emerson Papers. Department of Entomology, American Museum of Natural History, New York, New York.
CEMP	Charles E. Merriam Papers. Joseph Regenstein Library, University of Chicago, Chicago, Illinois.
EBC	Edwin M. Banks Collection. University of Illinois Archives, Urbana, Illinois.
■	Faculty Minute Papers. Lake Forest College, Lake Forest, Illinois.
FRLP-MBL	Frank Rattray Lillie Papers. Library of the Marine Biological Laboratory, Woods Hole, Massachusetts.
FRLP-UC	Frank Rattray Lillie Papers. Joseph Regenstein Library, University of Chicago, Chicago, Illinois.
■	Ernst Mayr. Correspondence. Harvard University Archives, Cambridge, Massachusetts.
GKNP	Gladwyn Kingsley Noble Papers. Department of Herpetology, American Museum of Natural History, New York, New York.
KPSP	Karl Patterson Schmidt Papers. Field Museum of Natural History Library, Chicago, Illinois.
RFA	Rockefeller Foundation Archives. Rockefeller Archive Center, North Tarrytown, New York.
UCPP	University of Chicago Presidents' Papers. Joseph Regenstein Library, University of Chicago, Chicago, Illinois.
WCAP	Warder Clyde Allee Papers. Joseph Regenstein Library, University of Chicago, Chicago, Illinois.

■ Woman's Peace Party Papers. Swarthmore College Peace Collection, Swarthmore, Pennsylvania.

ZDR Zoology Department Records. Joseph Regenstein Library, University of Chicago, Chicago, Illinois.

Published Sources

Aberle, Sophie D. and George W. Corner. *Twenty-Five Years of Sex Research: History of the National Research Committee for Research in Problems of Sex, 1922–1947*. Philadelphia: W. B. Saunders Co., 1953.

Adams, Charles C. "Baseleveling and Its Faunal Significance, with Illustrations from the Southeastern United States." *American Naturalist* 35 (1901): 839–51.

———. *Guide to the Study of Animal Ecology*. New York: Macmillan Co., 1913.

———. "The Postglacial Dispersal of the North American Biota." *Biological Bulletin* 9 (1905): 53–71.

———. "Southeastern United States as a Center of Geographical Distribution of Flora and Fauna." *Biological Bulletin* 3 (1902): 115–31.

———. "The Variations and Ecological Distribution of the Snails of the Genus Io." *Memoirs of the National Academy of Sciences* 12 (1915): 1–184.

Adams, Charles C., and G. D. Fuller. "Henry Chandler Cowles, Physiographic Plant Ecologist." *Annals of the Association of American Geographers* 30 (1940): 39–43.

Addams, Jane. *Peace and Bread*. New York: Macmillan Co., 1922.

Adler, Mortimer J. "The Chicago School." *Harper's* 183 (1941): 383–84.

Allee, Marjorie Hill. *The Camp at Westlands*. Boston: Houghton Mifflin Co., 1941.

———. *The Great Tradition*. Boston: Houghton Mifflin Co., 1937.

———. "History of the Chicago Meeting of Orthodox Friends." 57th Street Meeting of Friends, Chicago, Ill. Typescript.

———. *The House*. Cambridge, Mass.: Riverside Press, 1944.

———. *Jane's Island*. Boston: Houghton Mifflin Co., 1931.

———. "Recipe for a Book." *Illinois Libraries* 20 (1938): 3–10.

Allee, Warder C. "About Warder Clyde Allee." In *What is Science?*, edited by J. R. Newman, 228–30. New York: Simon & Schuster, 1955.

———. "Animal Aggregations." *Anatomical Record* 17 (1920): 340.

———. "Animal Aggregations." *Quarterly Review of Biology* 2 (1927): 367–98.

———. "Animal Aggregations: A Request for Information." *Condor* 25 (1923): 129–31.

———. *Animal Aggregations: A Study in General Sociology*. Chicago: University of Chicago Press, 1931.

———. *Animal Life and Social Growth*. Baltimore: Williams & Wilkins Co., 1932.

———. "Biology and International Relations." *New Republic* 112 (1945): 816–17.
———. "The Biology of Disarmament." *Quaker* 2 (1921): 173–74.
———. "The Biology of Peace." *Quaker* 1 (1920): 39–41.
———. "Certain Relations between Rheotaxis and Resistance to Potassium Cyanide in Isopoda." *Journal of Experimental Zoology* 16 (1914): 397–412.
———. "Chemical Control of Rheotaxis in Asellus." *Journal of Experimental Zoology* 21 (1916): 163–98.
———. "Concerning the Origin of Sociality in Animals." *Scientia* 67 (1940): 154–60.
———. "Cooperation among Animals." *Chicago Alumni Magazine* 20 (1928): 418–25.
———. "Cooperation and Conflict as Modes of Social Integration." Lecture delivered at the Oriental Institute, Chicago, Illinois, 7 November 1945.
———. "Distribution of Animals in a Tropical Rain-Forest with Relation to Environmental Factors." *Ecology* 7 (1926): 445–68.
———. "Dominance and Hierarchy in Societies of Vertebrates." *Structure et Physiologie des Sociétés Animales.* Colloques Internationaux du Centre National de la Recherche Scientifique 34 (1950–52): 151–81.
———. "The Ecological Importance of the Rheotactic Reaction of Isopods." *Biological Bulletin* 27 (1914): 52–66.
———. "The Effect of Bunching on Oxygen Consumption in Land Isopods." *Anatomical Record* 31 (1925): 336.
———. "An Experimental Analysis of the Relation between Physiological States and Rheotaxis in Isopoda." *Journal of Experimental Zoology* 13 (1912): 270–344.
———. "Further Studies on Physiological States and Rheotaxis in Isopoda." *Journal of Experimental Zoology* 15 (1913): 257–95.
———. "Human Conflict and Co-operation: The Biological Background." In *Approaches to National Unity, Fifth Symposium on the Conference on Science, Philosophy and Religion,* edited by L. Bryson, L. Finkelstein, and R. M. MacIver, 321–67. New York: Harper & Bros., 1945.
———. "Integration of Problems concerning Protozoan Populations with Those of General Biology." *American Naturalist* 75 (1941): 473–87.
———. "Measurement of Environmental Factors in the Tropical Rain-Forest of Panama." *Ecology* 7 (1926): 273–303.
———. "Science Confirms an Old Faith." *American Friend* 35 (1928): 780–82, 798–99.
———. *The Social Life of Animals.* New York: W. W. Norton & Co., 1938.
———. "Studies in Animal Aggregations: Causes and Effects of Bunching in Land Isopods." *Journal of Experimental Zoology* 45 (1926): 255–77.
———. "Studies in Animal Aggregations: Mass Protection from Fresh Water for Procerodes, a Marine Turbellarian." *Journal of Experimental Zoology* 50 (1928): 295–318.
———. "Studies in Animal Aggregations: Mass Protection From Hypotonic

Sea-Water for Procerodes, a Marine Turbellarian, with Total Electrolytes Controlled." *Journal of Experimental Zoology* 54 (1929): 349–81.

———. "Studies in Animal Aggregations: Some Physiological Effects of Aggregation on the Brittle Starfish, Ophioderma Brevispina." *Journal of Experimental Zoology* 48 (1927): 475–95.

———. "Studies in Marine Ecology: I. The Distribution of Common Littoral Invertebrates of the Woods Hole Region." *Biological Bulletin* 44 (1923): 167–91.

———. "Studies in Marine Ecology: III. Some Physiological Factors Related to the Distribution of Littoral Invertebrates." *Biological Bulletin* 44 (1923): 205–53.

———. "Studies in Marine Ecology: IV. The Effect of Temperature in Limiting the Geographic Range of Invertebrates of the Woods Hole Littoral." *Ecology* 4 (1923): 341–54.

———. "Where Angels Fear to Tread: A Contribution from General Sociology to Human Ethics." *Science* 97 (1943): 517–25.

Allee, W. C., and M. H. Allee. "Adventures of a College Professor." *American Friend* 7 (1919): 1132–33.

———. *Jungle Island*. Chicago: Rand McNally & Co., 1925.

Allee, W. C., Mary N. Allee, Frances Ritchey, and Elizabeth W. Castles. "Leadership in a Flock of White Pekin Ducks." *Ecology* 28 (1947): 310–15.

Allee, W. C., E. S. Bowen, J. C. Welty, and R. Oesting. "The Effect of Homotypic Conditioning of Water on the Growth of Fishes, and Chemical Studies of the Factors Involved." *Journal of Experimental Zoology* 68 (1934): 183–213.

Allee, W. C., and N. E. Collias. "The Effect of Estradiol on the Social Organization of Flocks of Hens." *Endocrinology* 27 (1940): 87–94.

Allee, W. C., N. E. Collias, and Elizabeth Beeman. "The Effect of Thyroxin on the Social Order in Flocks of Hens." *Endocrinology* 27 (1940): 827–35.

Allee, W. C., N. E. Collias, and Catherine Z. Lutherman. "Modification of the Social Order in Flocks of Hens by the Injection of Testosterone Proprionate." *Physiological Zoology* 12 (1939): 412–40.

Allee, W. C., A. E. Emerson, O. Park, T. Park, and K. P. Schmidt. *The Principles of Animal Ecology*. Philadelphia: W. B. Saunders, 1949.

Allee, W. C., A. J. Finkel, and W. H. Hoskins. "The Growth of Goldfish in Homotypically Conditioned Water; A Population Study in Mass Physiology." *Journal of Experimental Zoology* 84 (1940): 417–43.

Allee, W. C., Peter Frank, and Marjorie Berman. "Homotypic and Heterotypic Conditioning in Relation to Survival and Growth of Certain Fishes." *Physiological Zoology* 19 (1946): 243–58.

Allee, W. C., and H. T. Odum. "A Note on the Stable Point of Populations Showing Both Intraspecific Cooperation and Disoperation." *Ecology* 35 (1954): 95–97.

Allee, W. C., R. B. Oesting, and W. H. Hoskins. "Is Food the Effective

Growth-Promoting Factor in Homotypically Conditioned Water?" *Physiological Zoology* 9 (1936): 409–32.

Allee, W. C., and Thomas Park. "Concerning Ecological Principles." *Science* 89 (1939): 166–69.

Allee, W. C., and G. M. Rosenthal. "Group Survival Value for *Philodina roseola*, a Rotifer." *Ecology* 30 (1949): 395–97.

Allee, W. C., and J. F. Schuett. "Studies in Animal Aggregations: The Relationship between Mass of Animals and Resistance to Colloidal Silver." *Biological Bulletin* 53 (1927): 301–17.

Allee, W. C., and E. R. Stein, Jr. "Light Reactions and Metabolism in May-Fly Nymphs." *Journal of Experimental Zoology* 26 (1918): 423–58.

Allen, Garland E. "The Eugenics Record Office at Cold Spring Harbor: An Essay in Institutional History." *Osiris*, 2d. ser., vol. 2 (1986): 225–64.

———. *Life Science in the Twentieth Century.* Cambridge: Cambridge University Press, 1978.

———. "Old Wine in New Bottles: From Eugenics to Population Control in the Work of Raymond Pearl." In *The Expansion of American Biology*, edited by Keith R. Benson, Ronald Rainger, and Jane Maienschein, 231–61. New Brunswick, N.J.: Rutgers University Press, 1991.

———. "Thomas Hunt Morgan and the Problem of Natural Selection." *Journal of the History of Biology* 1 (1968): 113–39.

———. *Thomas Hunt Morgan: The Man and His Science.* Princeton, N.J.: Princeton University Press, 1978.

Anderson, H. H. "Domination and Integration in the Social Behavior of Young Children in An Experimental Play Situation." *Genetic Psychology Monographs* 19 (1937): 341–408.

———. "Domination and Social Integration in the Behavior of Kindergarten Children in an Experimental Play Situation." *Journal of Experimental Education* 8 (1939): 123–31.

———. "An Examination of the Concepts of Domination and Integration in Relation to Dominance and Ascendance." *Psychology Review* 47 (1940): 21–37.

———. "An Experimental Study of Dominative and Integrative Behavior in Children of Preschool Age." *Journal of Social Psychology* 8 (1937): 335–45.

———. "Studies in Dominative and Socially Integrative Behavior." *American Journal of Orthopsychiatry* 15 (1945): 133–39.

Anshen, Ruth Nanda. *Freedom: Its Meaning.* New York: Harcourt, Brace, 1940.

Arbib, Michael A., and Mary B. Hesse. *The Construction of Reality.* Cambridge: Cambridge University Press, 1986.

Ardrey, Robert. *The Territorial Imperative: A Personal Inquiry into the Animal Origins of Property and Nations.* New York: Athaneum, 1966.

Augros, Robert, and George Stanciu. *The New Biology: Discovering the Wisdom of Nature.* Boston: Shambhala, 1987.

Banks, Edwin M. "Warder Clyde Allee and the Chicago School of Animal Behavior." *Journal of the History of the Behavioral Sciences* 21 (1985): 345–53.

Bannister, Robert C. *Social Darwinism: Science and Myth in Anglo-American Social Thought*. Philadelphia: Temple University Press, 1979.

Barrows, Harlan H. "Geography as Human Ecology." *Annals of the Association of American Geographers* 13 (1923): 1–14.

Beach, Frank A. "Historical Origins of Modern Research on Hormones and Behavior." *Hormones and Behavior* 15 (1981): 325–76.

Beatty, John. "Dobzhansky and the Significance of Genetic Load." Typescript.

———. "Ecology and Evolutionary Biology in the War and Postwar Years: Questions and Comments." *Journal of the History of Biology* 21 (1988): 245–63.

Beebe, William, G. Innes Hartley, and Paul G. Howes. *Tropical Wild Life in British Guiana; Zoological Contributions from the New York Zoological Society*. New York: New York Zoological Society, 1917.

Beeman, Elizabeth A., and W. C. Allee. "Some Effects of Thiamin on the Winning of Social Contacts in Mice." *Physiological Zoology* 18 (1945): 195–221.

Bellomy, Donald C. "'Social Darwinism' Revisited." *Perspectives in American History* n. s. 1 (1984): 1–129.

Bigelow, Maurice H. "Contributions of Zoology to Human Welfare." *Science* 48 (1918): 1–5.

Birke, Lynda. *Women, Feminism and Biology: The Feminist Challenge*. Brighton: Harvester Press, 1986.

Blake, Lincoln. "The Concept and Development of Science at the University of Chicago, 1890–1905." Ph.D. diss., University of Chicago, 1977.

Bleier, Ruth. *Science and Gender: A Critique of Biology and Its Theories on Women*. New York: Pergamon Press, 1984.

Bodenheimer, F. S. *Problems in Animal Ecology*. London: Oxford University Press, 1938.

Bohn, Georges, and A. Drzewina. "Variations de la sensibilité à l'eau douce des convoluta, suivant les états physiologiques et le nombre des animaux en expérience." *Comptes rendus de l'Académie des Sciences* 171 (1920): 1023–25.

Borgese, Elisabeth M. "The Animal Kingdom and the World Republic." *Common Cause* 1 (1947): 75–76.

Botkin, Daniel. *Discordant Harmonies: A New Ecology for the Twenty-First Century*. New York: Oxford University Press, 1990.

Bowler, Peter J. *The Eclipse of Darwinism: Anti-Darwinian Evolution Theories in the Decades around 1900*. Baltimore: Johns Hopkins University Press, 1983.

———. "Hugo De Vries and Thomas Hunt Morgan: The Mutation Theory and the Spirit of Darwinism." *Annals of Science* 35 (1978): 55–73.

Brandon, Robert N., and Richard M. Burian, eds. *Genes, Organisms, Populations*. Cambridge, Mass.: MIT Press, 1984.

Braun-Blanquet, J. *Plant Sociology: The Study of Plant Communities.* Translated by George D. Fuller and Henry S. Conrad. New York: McGraw-Hill Book Co., 1932.

Bresslau, E. "Die Ausscheidung von Schutzstoffen bei einzelligen Lebewesen." *Bericht der Senckenbergischen Naturforschender Gesellschaft* 54 (1924): 49–67.

Brody, Samuel. "Science and Social Wisdom." *Scientific Monthly* 59 (1944): 203–14.

Brown, D. M. "Ecological Concepts and Human Welfare." *Journal of the Tennessee Academy of Science* 18 (1943): 142–48.

Bulmer, Martin. *The Chicago School of Sociology: Institutionalization, Diversity, and the Rise of Sociological Research.* Chicago: University of Chicago Press, 1984.

———. "The Early Institutional Establishment of Social Science Research: The Local Community Research Committee at the University of Chicago, 1923–30." *Minerva* 18 (1980): 50–110.

Burkhardt, Richard W., Jr. "Charles Otis Whitman, Wallace Craig, and the Biological Study of Animal Behavior in the United States, 1898–1925." In *The American Development of Biology,* edited by Ronald Rainger, Keith R. Benson, and Jane Maienschein, 185–218. Philadelphia: University of Pennsylvania Press, 1988.

———. "The Development of an Evolutionary Ethology." In *Evolution from Molecules to Men,* edited by D. S. Bendall, 429–44. Cambridge: Cambridge University Press, 1983.

———. "On the Emergence of Ethology as a Scientific Discipline." *Conspectus of History* 1 (1981): 62–81.

Burns, Edward McNall. *David Starr Jordan: Prophet of Freedom.* Stanford, Calif.: Stanford University Press, 1953.

Cain, Arthur J., and Philip M. Sheppard. "The Theory of Adaptive Polymorphism." *American Naturalist* 88 (1954): 321–26.

Cannon, Walter B. "Physiological Regulation of Normal States: Some Tentative Postulates concerning Biological Homeostatics." In *À Charles Richet,* edited by Auguste Pettit, 91–93. Paris: Éditions Médicales, 1926.

———. *The Wisdom of the Body.* New York: W. W. Norton & Co., 1932.

Cantor, Geoffrey N. "Weighing Light: The Role of Metaphor in Eighteenth-Century Optical Discourse." In *The Figural and the Literal: Problems of Language in the History of Science and Philosophy, 1630–1800,* edited by A. E. Benjamin, G. N. Cantor, and J. R. R. Christie, 124–46. Manchester: Manchester University Press, 1987.

Caplan, Arthur L. "Ethics, Evolution, and the Milk of Human Kindness." In *The Sociobiology Debate: Readings on the Ethical and Scientific Issues concerning Sociobiology,* edited by Arthur L. Caplan, 304–15. New York: Harper & Row, 1978.

Caron, Joseph A. "La Théorie de la coopération animale dans l'écologie de W. C. Allee: Analyse du double registre d'un discours." Master's thesis, University of Montreal, 1977.

Carpenter, Clarence R. "A Field Study of the Behavior and Social Relations of Howling Monkeys (Alouatta Palliata)." *Comparative Psychology Monographs* 10 (1934): 1–168.

Carr, Harvey R., ed. *Posthumous Works of Charles Otis Whitman*. vol. 3. *The Behavior of Pigeons*. Washington, D.C.: Carnegie Institution, 1919.

Carr-Saunders, Alexander M. *The Population Problem: A Study in Human Evolution*. Oxford: Clarendon Press, 1922.

Castle, Gordon B. "The Experimental Determination of Caste Differentiation in Termites." *Science* 80 (1934): 314.

———. "An Experimental Investigation of Caste Differentiation in *Zootermopsis angusticollis*." In *Termites and Termite Control*. 2d ed., edited by C. A. Kofoid, 292–310. Berkeley: University of California Press, 1934.

Chamberlin, Rollin T. "Biographical Memoir of Thomas C. Chamberlin, 1843–1928." *National Academy of Sciences Biographical Memoirs* (1934): 307–407.

———. "The Origin and History of the Earth." In *The World and Man as Science Sees Them*, edited by Forest R. Moulton, 48–98. Chicago: University of Chicago Press, 1937.

Chamberlin, Thomas C., and Rollin D. Salisbury. *A College Text-book of Geology*. New York: Henry Holt & Co., 1909.

———. *A General Treatise on Geology*. 3 vols. New York: Henry Holt & Co., 1906.

Chapman, Royal N. *Animal Ecology with Especial Reference to Insects*. New York: McGraw-Hill, 1931.

———. "The Causes of Fluctuation of Populations of Insects." *Proceedings of the Hawaiian Entomological Society* 8 (1933): 279–92.

———. "An Experimental Analysis of the Cause of Population Fluctuations." *Science* 80 (1934): 297–98.

———. "The Quantitative Analysis of Environmental Factors." *Ecology* 9 (1928): 111–22.

Chatfield, Charles. *For Peace and Justice: Pacifism in America, 1910–1941*. Knoxville: University of Tennessee Press, 1971.

———. "World War I and the Liberal Pacifist in the United States." *American Historical Review* 75 (1970): 1920–37.

Child, Charles M. "Behavior Origins from a Physiologic Point of View." *American Medical Association Archives of Neurology and Psychiatry* 15 (1926): 173–84.

———. *Patterns and Problems of Development*. Chicago: University of Chicago Press, 1941.

———. *Physiological Foundations of Behavior*. New York: Henry Holt & Co., 1924.

———. "The Process of Reproduction in Organisms." *Biological Bulletin* 23 (1912): 1–39.

———. "The Relation between Functional Regulation and Form Regulation." *Journal of Experimental Zoology* 3 (1906): 559–82.

———. "Social Integration as a Biological Process." *American Naturalist* 74 (1940): 389–97.

———. "Some Considerations Regarding So-Called Formative Substances." *Biological Bulletin* 11 (1906): 165–81.

Churchill, Frederick B. "August Weismann and a Break from Tradition." *Journal of the History of Biology* 1 (1968): 91–112.

Cittadino, Eugene. "Ecology and the Professionalization of Botany in America, 1890–1905." *Studies in the History of Biology* 4 (1980): 171–98.

Clark, Terry N., ed. *Gabriel Tarde. On Communication and Social Influence.* Chicago: University of Chicago Press, 1969.

Clarke, Adele E. "Emergence of the Reproductive Research Enterprise: A Sociology of Biological, Medical, and Agricultural Science in the United States, 1910–1940." Ph.D. diss., University of California, San Francisco, 1985.

———. "Research Materials and Reproductive Science in the United States, 1910–1940." In *Physiology in the American Context, 1850–1940*, edited by Gerald L. Geison, 323–50. Bethesda, Md.: American Physiological Society, 1987.

Clements, Frederic E., and V. E. Shelford. *Bio-Ecology.* New York: John Wiley & Sons, 1939.

Coker, Robert E. "What Are the Fittest? I. A Mischievous Fallacy." *Scientific Monthly* 55 (1942): 487–494.

Cole, Leon J. "Biological Philosophy and the War." *Scientific Monthly* 8 (1918): 247–57.

Coleman, William. "Evolution into Ecology? The Strategy of Warming's Ecological Plant Geography." *Journal of the History of Biology* 19 (1986): 181–96.

———. "Science and Symbol in the Turner Frontier Hypothesis." *American Historical Review* 72 (1967): 22–49.

Collias, N. E. "Aggressive Behavior among Vertebrate Animals." *Physiological Zoology* 17 (1944): 83–123.

———. "The Role of American Zoologists and Behavioural Ecologists in the Development of Animal Sociology, 1934–1964." *Animal Behaviour* 41 (1991): 613–32.

Collie, George L., and Hiram D. Densmore. *Thomas C. Chamberlin, Ph.D., Sc.D., LL.D. and Rollin D. Salisbury, LL.D.: A Beloit College Partnership.* Evansville, Wis.: Antes Press, 1932.

Collins, James P. "*Evolutionary Ecology* and the Use of Natural Selection in Ecological Theory." *Journal of the History of Biology* 19 (1986): 257–88.

Conklin, Edwin Grant. "Biology and Democracy." *Scribner's* 65 (1919): 403–12.

———. *The Direction of Human Evolution.* Rev. ed. New York: Charles Scribner's Sons, 1934.

———. "Does Science Afford a Basis for Ethics?" *Scientific Monthly* 49 (1939): 295–303.

———. *Man, Real and Ideal.* New York: Charles Scribner's Sons, 1943.

———. "Science and Ethics." *Vital Speeches of the Day* 4 (1938): 194–200.

Cook, Robert Edward. "Raymond Lindeman and the Trophic-Dynamic Concept in Ecology." *Science* 198 (1977): 22–26.

Cook, Sarah Gibbard. "Cowles Bog, Indiana, and Henry Chandler Cowles." Chesterton, Ind.: Indiana Dunes National Lakeshore, 1980.

Cooper, William S. "Henry Chandler Cowles." *Ecology* 16 (1935): 281–83.

Coulter, John Merle. *Plant Relations: A First Book in Botany.* New York: D. Appleton, 1900.

Cowles, Henry Chandler. "The Causes of Vegetative Cycles." *Botanical Gazette* 51 (1911): 161–83.

———. "The Ecological Relations of the Vegetation on the Sand Dunes of Lake Michigan." *Botanical Gazette* 27 (1899): 95–117, 167–202, 281–308, 361–91.

———. "The Physiographic Ecology of Chicago and Vicinity; A Study of the Origin, Development, and Classification of Plant Societies." *Botanical Gazette* 31 (1901): 73–108, 145–82.

———. "The Plant Societies of Chicago and Vicinity." *Bulletin of the Geographic Society of Chicago* 2 (1901): 1–76.

Cox, David Lee. "Charles Elton and the Emergence of Modern Ecology." Ph.D. diss., Washington University, St. Louis, 1979.

Craig, Wallace. "Expression of the Emotions in the Pigeons." Ph.D. diss., University of Chicago, 1908.

———. "North Dakota Life: Plant, Animal and Human." *Bulletin of the American Geographical Society* 40 (1908): 321–415.

Crawford, Meredith P. "The Social Psychology of the Vertebrates." *Psychological Bulletin* 36 (1939): 407–46.

Crook, David Paul. "Nature's Pruning Hook? War and Evolution, 1890–1918: A Response to Nancy Stepan." *Australian Journal of Politics and History* 33 (1987): 238–51.

———. "Peter Chalmers Mitchell and Antiwar Evolutionism in Britain during the Great War." *Journal of the History of Biology* 22 (1989): 325–56.

Cross, Stephen J., and William R. Albury. "Walter B. Cannon, L. J. Henderson, and the Organic Analogy." *Osiris,* 2d ser., vol. 3 (1987): 165–92.

Crowcroft, Peter. *Elton's Ecologists: A History of the Bureau of Animal Population.* Chicago: University of Chicago Press, 1991.

Crunden, Robert M. *From Self to Society, 1919–1941.* Englewood Cliffs, N.J.: Prentice-Hall, 1972.

———. *Ministers of Reform: The Progressives' Achievement in American Civilization, 1889–1920.* New York: Basic Books, 1982.

———. *Progressivism,* edited by John D. Buenker, John C. Burnham, and Robert M. Crunden, 71–104. Cambridge, Mass.: Schenkmann Publishing Co., 1977.

Cutler, D. W., and L. M. Crump. "The Influence of Washing on the Reproductive Rate of *Colpidium colpoda.* II." *Biochemical Journal* 19 (1925): 450–53.

———. "The Rate of Reproduction in Artificial Cultures of *Colpidium colpoda*. I." *Biochemical Journal* 17 (1923): 175–86.
———. "The Rate of Reproduction in Artificial Cultures of *Colpidium colpoda*. II." *Biochemical Journal* 17 (1923): 875–86.
———. "The Rate of Reproduction in Artificial Cultures of *Colpidium colpoda*. III." *Biochemical Journal* 18 (1924): 903–12.
Dahl, Robert A. *A Preface to Democratic Theory.* Chicago: University of Chicago Press, 1956.
Darden, Lindley. "William Bateson and the Promise of Mendelism." *Journal of the History of Biology* 10 (1977): 87–106.
Darling, F. Fraser. *A Herd of Red Deer.* Oxford: Clarendon Press, 1937.
Darwin, Charles. *On the Origin of Species.* 1859. Facsimile of 1st ed. Cambridge, Mass.: Harvard University Press, 1964.
Davenport, Charles B. "The Animal Ecology of the Cold Spring Sand Spit, with Remarks on the Theory of Adaptation." *Decennial Publications of the University of Chicago* 10 (1903): 157–76.
———. "The Collembola of Cold Spring Beach, with Special Reference to the Movements of the Poduridae." *Cold Spring Harbor Monographs* 2 (1903): 3–30.
———. *Experimental Morphology.* 2d ed. 2 vols. in 1. New York: Macmillan Co., 1908.
Davis, David E. "An Inquiry into the Phylogeny of Gangs." In *Roots of Behavior,* edited by Eugene Bliss, 316–22. New York: Harper & Bros., 1962.
DeBenedetti, Charles. *Origins of the Modern American Peace Movement, 1915–1929.* Millwood, N.Y.: KTO Press, 1978.
———. *The Peace Reform in American History.* Bloomington: Indiana University Press, 1980.
———., ed. *Peace Heroes in the Twentieth Century.* Bloomington: Indiana University Press, 1986.
Deegener, Paul Johannes. *Die Formen der Vergesellschaftung in Tierreiche. Ein systematisch-soziologischer Versuch.* Leipzig: Verlag von Veit & Comp., 1918.
Dewey, John. "Antinaturalism in Extremis." In *Naturalism and the Human Spirit,* edited by Yervant H. Krikorian, 1–16. New York: Columbia University Press, 1944.
———. *Democracy and Education.* 1916. Reprint. New York: Macmillan Co., 1961.
———. "The Reflex Arc Concept in Psychology" (1896). In *The Early Works of John Dewey, 1882–1898,* edited by Jo Anne Boydston. Vol. 5: 96–110. Carbondale: Southern Illinois University Press, 1972.
———. "The Relation of Philosophy to Theology." In *The Early Works of John Dewey: 1882–1898,* edited by Jo Anne Boydston. Vol. 4. Carbondale: Southern Illinois University Press, 1969.
Dewsbury, Donald A., ed. *Leaders in the Study of Animal Behavior.* Lewisburg, Pa.: Bucknell University Press, 1985.

Dobzhansky, Theodosius. *Genetics and the Origin of Species.* 3d ed. New York: Columbia University Press, 1951.

Dollard, John, L. W. Doob, N. E. Miller, O. H. Mowrer, and R. R. Sears. *Frustration and Aggression.* New Haven, Conn.: Yale University Press, 1939.

Drzewina, Anna, and Georges Bohn. "Action nocive de l'eau sur des stentors, en fonction de la masse du liquide." *Comptes rendus de la Societé de Biologie* 84 (1921): 917–20.

———. "La défense des animaux groupés vis-à-vis agents nocifs." *Comptes rendus de l'Académie des Sciences* 172 (1921): 779–81.

———. "Sur des phénomènes d'auto-destruction et d'auto-agglutination chez les Convoluta." *Comptes rendus de l'Académie des Sciences* 174 (1922): 330–32.

———. "Sur des phénomènes d'auto-protection et d'auto-destruction chez des animaux aquatiques." *Comptes rendus de l'Académie des Sciences* 173 (1921): 107–9.

———. "Variations dans le temps de la résistance aux agents physiques et chimiques, chez *Rana fusca.*" *Comptes rendus de la Societé de Biologie* 84 (1921): 963–65.

———. "Variations de la susceptibilité aux agents nocifs avec le nombre des animaux Traités." *Comptes rendus de l'Académie des Sciences* 172 (1921): 485–87.

Dublin, Louis I., ed. *Population Problems in the United States and Canada.* Boston: Houghton Mifflin Co., 1926.

Durant, John R. "The Beast in Man: An Historical Perspective on the Biology of Human Aggression." In *The Biology of Aggression,* edited by Paul F. Brain and David Benton, 17–46. Rockville, Md.: Sitjhoof & Noordhoff, 1981.

———. "Innate Character in Animals and Man: A Perspective on the Origins of Ethology." In *Biology, Medicine and Society, 1840–1940,* edited by C. Webster, 157–92. Cambridge: Cambridge University Press, 1981.

Elton, Charles. *Animal Ecology.* London: Sidgwick & Jackson, 1927.

Emerson, Alfred E. "Basic Comparisons of Human and Insect Societies." *Biological Symposia* 8 (1942): 163–76.

———. "The Biological Basis of Social Cooperation." *Illinois Academy of Science Transactions* 39 (1946): 9–18.

———. "The Biological Foundations of Ethics and Social Progress." In *Goals of Economic Life,* edited by A. Dudley Ward, 277–304. New York: Harper & Bros., 1953.

———. "Biological Sociology." *Denison University Bulletin, Journal of Science Laboratories* 36 (1941): 146–55.

———. "Biological Species." *Encyclopaedia Britannica.* Chicago: Encyclopaedia Britannica, 1955.

———. "Cooperation and Conflict in the Balance of Nature." Lecture delivered at the Oriental Institute, Chicago, Illinois, November 1945.

———. "Development of a Soldier of *Nasutitermes Constrictotermes cavifrons* (Holmgren) and Its Phylogenetic Significance." *Zoologica* 7 (1926): 69–100.

———. "Ecology, Evolution, and Society." *American Naturalist* 77 (1943): 97–118.

———. "Ethospecies, Ethotypes, Taxonomy, and Evolution of *Apicotermes* and *Allognathotermes* (Isoptera, Termitidae)." *American Museum Novitates* no. 1771 (1956): 1–31.

———. "Human Cultural Evolution and Its Relation to Organic Evolution of Insect Societies." In *Social Change in Developing Areas: A Reinterpretation of Evolutionary Theory,* edited by H. R. Barringer, G. I. Blanksten, and R. W. Mack, 50–67. Cambridge, Mass.: Schenkman Publications Co., 1965.

———. "K. P. Schmidt—Herpetologist, Ecologist, Zoogeographer." *Science* 127 (1958): 1162–63.

———. "The Origin of Species." *Ecology* 19 (1938): 152–54.

———. "Populations of Social Insects." *Ecological Monographs* 9 (1939): 287–300.

———. "Regenerative Behavior and Social Homeostasis of Termites." *Ecology* 37 (1956): 248–58.

———. "Social Coordination and the Superorganism," *American Midland Naturalist* 21 (1939): 182–209.

———. "Systematics and Speciation." *Ecology* 24 (1943): 412–13.

———. "Taxonomic Categories and Population Genetics." *Entomological News* 56 (1945): 14–19.

———. "Taxonomy and Ecology." *Ecology* 22 (1941): 213–15.

———. "Termite Nests—A Study of the Phylogeny of Behavior." *Ecological Monographs* 8 (1938): 249–84.

———. "Termitophile Distribution and Quantitative Characters as Indicators of Physiological Speciation in British Guiana Termites (Isoptera)." *Annals of the Entomological Society of America* 28 (1935): 369–95.

———. "Why Termites?" *Scientific Monthly* 64 (1947): 337–45.

Emerson, Alfred E., and Thomas Park. "Warder Clyde Allee: Ecologist and Ethologist." *Science* 121 (1955): 686–87.

Emlen, John T., Jr., and F. W. Lorenz. "Pairing Responses of Free-Living Quail to Sex Hormone Implants." *Auk* 59 (1942): 369–78.

Engel, J. Ronald. *Sacred Sands: The Struggle for Community in the Indiana Dunes.* Middletown, Conn.: Wesleyan University Press, 1983.

Espinas, Alfred. *Des Sociétés Animales.* 2d ed. Paris: Librairie Germer Baillière, 1878.

Evans, Gertrude. "The Relation between Vitamins and the Growth and Survival of Goldfishes in Homotypically Conditioned Water." *Journal of Experimental Zoology* 74 (1936): 449–76.

Evans, Mary Alice, and Howard Ensign Evans. *William Morton Wheeler, Biologist.* Cambridge, Mass.: Harvard University Press, 1970.

Fausto-Sterling, Anne. *Myths of Gender: Biological Theories about Women and Men.* New York: Basic Books, 1985.

Filene, Peter G. "An Obituary for 'The Progressive Movement.'" *American Quarterly* 22 (1970): 20–34.

Finkel, A. J., and W. C. Allee. "The Effect of Traces of Tin on the Rate of Growth of Goldfish." *American Journal of Physiology* 130 (1940): 665–70.

Finney, Jack. *The Body Snatchers*. New York: Dell Publishing Co., 1955.

Fisher, Ronald A. *The Genetical Theory of Natural Selection*. Oxford: Oxford University Press, 1930.

Flitcraft, Harold W., ed. *History of the 57th Meeting of Friends*. Chicago: 57th Street Meeting of Friends, 1957.

Forbes, John. *The Quaker Star under Seven Flags, 1917–1927*. Philadelphia: University of Pennsylvania Press, 1962.

Fosdick, Raymond B. "Science and the Moral Order." *Science* 93 (1941): 306–7.

———. *The Story of the Rockefeller Foundation*. New York: Harper & Bros., 1952.

Francis, Richard. "On the Relationship between Aggression and Social Dominance." *Ethology* 78 (1988): 223–37.

Friley, Charles Edwin. "Science and Human Values." *Bios* 13 (1942): 67–75.

Gasman, Daniel. *The Scientific Origins of National Socialism: Social Darwinism in Ernst Haeckel and the German Monist League*. New York: American Elsevier, 1971.

Gatewood, W. B. *Controversy in the Twenties: Fundamentalism, Modernism, and Evolution*. Nashville, Tenn.: Vanderbilt University Press, 1969.

Gause, G. F. "The Principles of Biocoenology." *Quarterly Review of Biology* 11 (1936): 320–36.

———. *The Struggle for Existence*. Baltimore: Williams & Wilkins, 1934.

Gerard, Ralph W. "A Biological Basis for Ethics." *Philosophy of Science* 9 (1942): 92–120.

———. "A Biologist's View of Society." *Common Cause* 3 (1949–50): 630–38.

———. "Cooperation and Conflict as Modes of Social Integration in Levels of Organization." Lecture delivered at the Oriental Institute, Chicago, Illinois, 7 November 1945.

———. "Higher Levels of Integration." In *Levels of Integration in Biological and Social Systems*, edited by Robert Redfield. Biological Symposia, vol. 8, pp. 67–88. Lancaster, Pa.: Jaques Cattell Press, 1942.

———. "Organic Freedom." In *Freedom: Its Meaning*, edited by Ruth Nanda Anshen, 412–27. New York: Harcourt, Brace & Co., 1940.

———. "Organism, Society and Science." *Scientific Monthly* 50 (1940): 340–50, 403–12, 530–35.

———. "Science and the Public." *Science* 106 (1947): 23–25.

Gerard, R. W., and A. E. Emerson. "Extrapolation from the Biological to the Social." *Science* 101 (1945): 582–585.

Gilbert, Scott. "Cellular Politics: Just, Goldschmidt, Waddington, and the Attempt to Reconcile Embryology and Genetics." In *The American*

Development of Biology, edited by Ronald Rainger, Keith R. Benson, and Jane Maienschein, 311–46. Philadelphia: University of Pennsylvania Press, 1988.

———. "The Embryological Origins of the Gene Theory." *Journal of the History of Biology* 11 (1978): 307–51.

Gilpin, M. *Group Selection in Predator-Prey Communities*. Princeton, N.J.: Princeton University Press, 1975.

Ginsburg, Benson, and W. C. Allee. "Some Effects of Conditioning on Social Dominance and Subordination in Inbred Strains of Mice." *Physiological Zoology* 15 (1942): 485–506.

Goode, J. Paul. "The Human Response to the Physical Environment." *Journal of Geography* (1904): 333–43.

Goodspeed, Thomas Wakefield. *A History of the University of Chicago*. Chicago: University of Chicago Press, 1916.

Gordon, Linda. "The Politics of Population: Birth Control and the Eugenics Movement." *Radical America* 8 (1974): 61–97.

———. *Woman's Body, Woman's Right: A Social History of Birth Control*. New York: Grossman's/Viking, 1976.

Gould, Stephen Jay. "The Hardening of the Modern Synthesis." In *Dimensions of Darwinism: Themes and Counterthemes in Twentieth-Century Evolutionary Theory*, edited by Marjorie Grene, 71–96. New York: Cambridge University Press, 1983.

———. *The Panda's Thumb: More Reflections in Natural History*. New York: W. W. Norton & Co., 1980.

Graebner, William. *The Engineering of Consent: Democracy and Authority in Twentieth-Century America*. Madison: University of Wisconsin Press, 1987.

Greenberg, Bernard. "Some Relations between Territory, Social Hierarchy, and Leadership in the Green Sunfish (*Lepomis Cyanellus*)." *Physiological Zoology* 20 (1947): 267–99.

Greene, John C. *Science, Ideology, and World View*. Berkeley and Los Angeles: University of California Press, 1981.

Greenleaf, W. E. "The Influence of Volume of Culture Medium and Cell Proximity on the Rate of Reproduction in Protozoa." *Proceedings of the Society of Experimental Biology and Medicine* 21 (1924): 405–6.

———. "The Influence of Volume of Culture Medium and Cell Proximity on the Rate of Reproduction of Infusoria." *Journal of Experimental Zoology* 46 (1926): 143–69.

Gregg, Robert E. "The Origin of Castes in Ants with Special Reference to *Pheidole Morrisi* Forel." *Ecology* 23 (1942): 295–308.

Grene, Marjorie, ed. *Dimensions of Darwinism: Themes and Counterthemes in Twentieth-Century Evolutionary Theory*. Cambridge: Cambridge University Press, 1983.

Griesemer, James R., and Michael J. Wade. "Laboratory Models, Causal Explanation and Group Selection." *Biology and Philosophy* 3 (1988): 67–96.

Griffen, Clyde. "The Progressive Ethos." In *The Development of an American Culture,* edited by Stanley Coben and Lorman Ratner. Englewood Cliffs, N.J.: Prentice-Hall, 1970.

Gruber, Carol S. *Mars and Minerva: World War I and the Uses of the Higher Learning in America.* Baton Rouge: Louisiana State University Press, 1975.

Guhl, A. M., and W. C. Allee. "Some Measurable Effects of Social Organization in Flocks of Hens." *Physiological Zoology* 17 (1944): 320–47.

Guillame, Ivan. "La Biosociologie d'Espinas et la sociobiologie de Wilson, deux systèmes de pensée comparables." *Cahiers Vilfredo Pareto* 23 (1985): 139–56.

Hagen, Joel B. "Ecologists and Taxonomists: Divergent Traditions in Twentieth-Century Plant Geography." *Journal of the History of Biology* 19 (1986): 197–214.

———. "Organism and Environment: Frederic Clement's Vision of a Unified Physiological Ecology." In *The American Development of Biology,* edited by Ronald Rainger, Keith R. Benson, and Jane Maienschein, 257–80. Philadelphia: University of Pennsylvania Press, 1988.

Hall, Diana Long. "Biology, Sex Hormones and Sexism in the 1920s." In *Women and Philosophy: Toward a Theory of Liberation,* edited by Carol C. Gould and Marx W. Wartofsky, 81–95. New York: Putnam, 1975.

Haller, Mark H. *Eugenics: Hereditarian Attitudes in American Thought.* New Brunswick, N.J.: Rutgers University Press, 1963.

Haraway, Donna J. "Animal Sociology and a Natural Economy of the Body Politic, Part I: A Political Physiology of Dominance." *Signs* 4 (1978): 21–36.

———. "The Biological Enterprise: Sex, Mind, and Profit from Human Engineering to Sociobiology." *Radical History Review* 20 (1979): 206–37.

———. *Crystals, Fabrics, and Fields.* New Haven, Conn.: Yale University Press, 1976.

———. "The High Cost of Information in Post–World War II Evolutionary Biology: Ergonomics, Semiotics, and the Sociobiology of Communication Systems." *Philosophical Forum* 13 (1981–82): 244–78.

———. *Primate Visions: Gender, Race, and Nature in the World of Modern Science.* New York: Routledge, Chapman & Hall, 1989.

———. "Signs of Dominance: From a Physiology to a Cybernetic System of Primate Society: C. R. Carpenter, 1930–1970." *Studies in the History of Biology* 6 (1983): 129–219.

———. "Teddy Bear Patriarchy: Taxidermy in the Garden of Eden, New York City, 1908–1936." *Social Text* 11 (1984/85): 20–64.

Harding, Sandra. *The Science Question in Feminism.* Ithaca, N.Y.: Cornell University Press, 1986.

Hare, Laura. "Caste Determination and Differentiation with Special Reference to the Genus Reticulitermes (Isoptera)." *Journal of Morphology* 56 (1934): 267–84.

Hecht, Susanna, and Alexander Cockburn. *The Fate of the Forest: Devel-*

opers, Destroyers, and Defenders of the Amazon. New York: Verso, 1989.

Helfand, Michael. "T. H. Huxley's 'Evolution and Ethics': The Politics of Evolution and the Evolution of Politics." *Victorian Studies* 20 (1977): 157–77.

Herrick, C. Judson. "A Neurologist Makes Up His Mind." *Scientific Monthly* 49 (1939): 99–110.

Hesse, Richard. *Ecological Animal Geography.* Translated by W. C. Allee and Karl P. Schmidt. New York: John Wiley & Sons, 1937.

Higham, John. "Hanging Together: Divergent Unities in American History." *Journal of American History* 61 (1974): 5–28.

Hirsch, Arnold R. *Making the Second Ghetto: Race and Housing in Chicago, 1940–1960*. Cambridge: Cambridge University Press, 1983.

Hofstadter, Richard. *Social Darwinism in American Thought*. Philadelphia: University of Pennsylvania Press, 1944.

Holmes, David R. *Stalking the Academic Communist: Intellectual Freedom and the Firing of Alex Novikoff*. Hanover, N.H.: University Press of New England, 1989.

Holmes, Samuel Jackson. "Darwinian Ethics and Its Practical Applications." *Science* 11 (1939): 117–23.

———. *The Evolution of Animal Intelligence*. New York: Henry Holt & Co., 1911.

———. *Life and Morals*. New York: Macmillan, 1948.

———. *Studies in Animal Behavior*. Boston: Richard G. Badger, 1916.

Hook, Sidney. "Naturalism and Democracy." In *Naturalism and the Human Spirit*, edited by Yervant H. Krikorian, 40–64. New York: Columbia University Press, 1944.

Howard, Walter E., and John T. Emlen, Jr. "Intercovey Social Relationships in the Valley Quail." *Wilson Bulletin* 54 (1942): 162–70.

Hughes, Helen M. "On Becoming a Sociologist." *Journal of the History of Sociology* 3 (1980–81): 27–39.

Hutchins, Robert M. "What Shall We Defend?" *Vital Speeches of the Day* 6 (1940): 547–49.

Hutchinson, G. E. "Bio-Ecology." *Ecology* 21 (1940): 267–68.

———. "Ecological Aspects of Succession in Natural Populations." *American Naturalist* 75 (1941): 406–18.

———. "Homage to Santa Rosalia, or Why Are There So Many Kinds of Animals?" *American Naturalist* 93 (1959): 137–45.

———. *The Kindly Fruits of the Earth: Reflections of an Embryo Ecologist*. New Haven, Conn.: Yale University Press, 1979.

———. "Limnological Studies in Connecticut. VII. A Critical Examination of the Supposed Relationship between Phytoplankton Periodicity and Chemical Changes in Lake Waters." *Ecology* 25 (1944): 3–26.

———. "A Note on the Theory of Competition between Two Social Species." *Ecology* 28 (1947): 319–21.

Hutchinson, G. E., and E. S. Deevey, Jr. "Ecological Studies on Populations." *Survey of Biological Progress* 1 (1949): 325–59.

Hutchinson, G. E., and Anne Wollack. "Studies on Connecticut Lake Sediments. II. Chemical Analyses of a Core from Linsley Pond, North Branford." *American Journal of Science* 238 (1940): 493–517.

Huxley, Julian S. "The Biologist Looks at Man." *Fortune* 36 (1942): 139–41, 146, 148, 150, 152.

———. *Touchstone for Ethics*. New York: Harper & Bros., 1947.

———. "War as a Biological Phenomenon." In *On Living in a Revolution*, 60–68. Edinburgh: T. & A. Constable, 1946.

Huxley, Thomas Henry. "Evolution and Ethics" (1893). In *Evolution and Ethics and Other Essays*. London: Macmillan, 1894.

Hyman, Libbie H. "Charles Manning Child, 1869–1954." *Biographical Memoirs of the National Academy of Sciences* 30 (1957): 73–103.

Jackson, Jeremy. "Interspecific Competition and Species' Distributions: The Ghosts of Theories and Data Past." *American Zoologist* 21 (1981): 889–901.

Jahn, T. L. "Studies on the Physiology of the Euglenoid Flagellates. I. The Relation of the Density of Population to the Growth Rate of *Euglena*." *Biological Bulletin* 57 (1929): 81–106.

Jennings, Herbert Spencer. *Behavior of the Lower Organisms*. New York: Columbia University Press, 1906.

———. "Raymond Pearl, 1879–1940." *Biographical Memoirs of the National Academy of Sciences* 22 (1942): 295–310.

Johannsen, William. "The Genotype Conception of Heredity." *American Naturalist* 45 (1911): 129–59.

Johnson, Willis H. "Effects of Population Density on the Rate of Reproduction in Oxytricha." *Physiological Zoology* 6 (1933): 22–54.

Jones, Gerald. *On Doing Good*. New York: Charles Scribner's Sons, 1971.

Jones, Lester M. *Quakers in Action*. New York: Macmillan Co., 1920.

Jones, Rufus M. *A Service of Love in Wartime*. New York: Macmillan Co., 1920.

Jordan, David Starr. *The Days of a Man*. 2 vols. Yonkers-on-Hudson, N.Y.: World Book Co., 1922.

———. *Democracy and World Relations*. Yonkers-on-the-Hudson, N.Y.: World Book Co., 1918.

———. *The Human Harvest*. Boston: Beacon Press, 1907.

———. *The Root of Evil: A Heart to Heart Talk of an Old Friend of the German People with His Fellow-Citizens of German Stock*. New York: American Friends of German Democracy, 1918.

———. "Social Darwinism." *Public* 21 (1918): 400–401.

———. "The Search for the Master Key of the Universe." *Scientific Monthly* 22 (1926): 33–36.

———. *War and the Breed*. Boston: Beacon Press, 1915.

———. "War and Genetic Values." *Journal of Heredity* 10 (1919): 223–25.

———. *War's Aftermath: A Preliminary Study of the Eugenics of War as Illustrated by the Civil War of the United States and the Late Wars in the Balkans.* Boston: Houghton Mifflin Co., 1914.

———. *Ways to Lasting Peace.* Indianapolis: Bobbs-Merrill Co., 1916.

Kaye, Howard L. *The Social Meaning of Modern Biology: From Social Darwinism to Sociobiology.* New Haven, Conn.: Yale University Press, 1986.

Keith, Sir Arthur. *Evolution and Ethics.* New York: G. B. Putnam's Sons, 1946.

Keller, Evelyn Fox. *Reflections on Gender and Science.* New Haven, Conn.: Yale University Press, 1985.

———. "Demarcating Public from Private Values in Evolutionary Discourse." *Journal of the History of Biology* 21 (1988): 195–211.

Kellogg, Vernon L. "The Bionomics of War: Race Modification by Military Selection." *Social Hygiene* 1 (1914–1915): 44–52.

———. *Darwinism Today.* New York: Henry Holt & Co., 1907.

———. "Eugenics and Militarism." *Atlantic Monthly,* 112 (1913): 99–108.

———. *Headquarters Nights.* Boston: Atlantic Monthly Press, 1917.

———. *Military Selection and Race Deterioration.* Oxford: Clarendon Press, 1916.

———. "War and Human Evolution: Germanized." *North American Review* 207 (1918): 364–69.

———. "War for Evolution's Sake." *Unpopular Review* 10 (1918): 146–59.

Kelly, Alfred. *The Descent of Darwin.* Chapel Hill: University of North Carolina Press, 1981.

Kendeigh, S. Charles. "Victor Ernest Shelford, Eminent Ecologist—1968." *Bulletin of the Ecological Society of America* 49 (1968): 97–100.

Kety, Seymour S. "Ralph Waldo Gerard, October 7, 1900–February 17, 1974." *Biographical Memoirs of the National Academy of Sciences* 53 (1982): 179–210.

Kevles, Daniel J. "George Ellery Hale, the First World War, and the Advancement of Science in America." *Isis* 59 (1968): 427–37.

———. *In the Name of Eugenics: Genetics and the Uses of Human Heredity.* Berkeley and Los Angeles: University of California Press, 1985.

———. *The Physicists: The History of a Scientific Community in Modern America.* 1977. Reprint. New York: Vintage Books, 1979.

Kimler, William C. "Advantage, Adaptiveness, and Evolutionary Ecology." *Journal of the History of Biology* 19 (1986): 215–33.

———. "Mimicry: Views of Naturalists and Ecologists before the Modern Synthesis." In *Dimensions of Darwinism: Themes and Counterthemes in Twentieth-Century Evolutionary Theory,* edited by Marjorie Grene, 97–128. Cambridge: Cambridge University Press, 1983.

Kingsland, Sharon E. *Modeling Nature: Episodes in the History of Population Ecology.* Chicago: University of Chicago Press, 1985.

———. "Toward a Natural History of the Human Psyche: C. M. Child, C. J. Herrick, and the Dynamic View of the Individual at the University of

Chicago." In *The Expansion of American Biology*, edited by Keith R. Benson, Ronald Rainger, and Jane Maienschein, 195–230. New Brunswick, N.J.: Rutgers University Press, 1991.

Kitcher, Philip. *Vaulting Ambition: Sociobiology and the Quest for Human Nature*. Cambridge, Mass.: MIT Press, 1985.

Kohler, Robert E. "A New Policy for the Patronage of Scientific Research: the Reorganization of the Rockefeller Foundation, 1921–1930." *Minerva* 14 (1976): 279–306.

———. "Warren Weaver and the Rockefeller Foundation Program in Molecular Biology: A Case Study in the Management of Science." In *The Sciences in the American Context: New Perspectives*, edited by Nathan Reinghold, 249–94. Washington, D.C.: Smithsonian Institution Press, 1979.

Kropotkin, Peter. "The Direct Action of Environment and Evolution." *Nineteenth Century and After* 85 (1919): 70–89.

———. "The Direct Action of the Environment on Plants." *Nineteenth Century and After* 68 (1910): 58–77.

———. "Inheritance of Acquired Characters." *Nineteenth Century and After* 71 (1912): 511–31.

———. *Memoirs of a Revolutionist*. 1899. Reprint. New York: Houghton Mifflin Co., 1930.

———. *Mutual Aid*. 1902. Reprint. Boston: Extending Horizons, 1955.

Kurtz, Lester R. *Evaluating Chicago Sociology*. Chicago: University of Chicago Press, 1984.

Kuznick, Peter J. *Beyond the Laboratory: Scientists as Political Activists in 1930s America*. Chicago: University of Chicago Press, 1987.

Kwa, Chunglin. "Mimicking Nature." Ph.D. diss., University of Amsterdam, 1989.

———. "Representations of Nature Mediating between Ecology and Science Policy: The Case of the International Biological Programme." *Social Studies of Science* 17 (1987): 413–42.

Lack, David. *The Natural Regulation of Animal Numbers*. Oxford: Clarendon Press, 1954.

Leake, Chauncey D. "Ethicogenesis." *Scientific Monthly* 60 (1945): 245–53.

Lerner, Gerda. *The Creation of Patriarchy*. New York: Oxford University Press, 1986.

Lerner, I. Michael, and Everett R. Dempster. "Indeterminism in Interspecific Competition." *Proceedings of the National Academy of Science* 48 (1962): 821–26.

Levine, Lawrence. *Defender of the Faith: William Jennings Bryan: The Last Decade, 1915–1925*. New York: Oxford University Press, 1965.

Lewin, Kurt, Ronald Lippitt, and Ralph K. White. "Patterns of Aggressive Behavior in Experimentally Created 'Social Climates.'" *Journal of Social Psychology*, 10 (1939): 271–99.

Lewis, J. David, and Richard L. Smith. *American Sociology and Pragmatism:*

Mead, Chicago, Sociology, and Symbolic Interaction. Chicago: University of Chicago Press, 1980.

Lewontin, R. C., Steven Rose, and Leon J. Kamin. *Not in Our Genes: Biology, Ideology, and Human Nature*. New York: Pantheon Books, 1984.

Lillie, Frank R. "Charles Otis Whitman." *Journal of Morphology* 22 (1911): xv–lxxvii.

———. "The Free-martin; a Study of the Action of Sex Hormones in the Foetal Life of Cattle." *Journal of Experimental Zoology* 23 (1917): 371–452.

———. "The Theory of Individual Development." *Popular Science Monthly* (September 1909): 239–52.

———. *The Woods Hole Marine Biological Laboratory*. Chicago: University of Chicago Press, 1944.

Lindeman, Raymond L. "The Trophic-Dynamic Aspect of Ecology." *Ecology* 23 (1942): 399–418.

Linton, Adelin, and Charles Wagley. *Ralph Linton*. New York: Columbia University Press, 1971.

Linton, Ralph. *The Study of Man: An Introduction*. New York: D. Appleton-Century Co., 1936.

Lippitt, Ronald. "An Experimental Study of the Effect of Democratic and Authoritarian Group Atmospheres." *University of Iowa Studies in Child Welfare* 16 (1940): 43–195.

———. "Field Theory and Experiment in Social Psychology: Autocratic and Democratic Group Atmospheres." *American Journal of Sociology* 45 (1939): 26–49.

Loeb, Jacques. "Biology and War." *Science* 45 (1917): 73–76.

———. *The Mechanistic Conception of Life*. Chicago: University of Chicago Press, 1912.

Lorenz, Konrad. *On Aggression*. Translated by Marjorie Kerr Wilson. New York: Harcourt Brace Jovanovich, 1966.

Ludmerer, Kenneth. *Genetics and American Society*. Baltimore: Johns Hopkins University Press, 1972.

Ludwig, William, and Charlotte Boost. "Uber das Wachstum von Protistenpopulationen and den allelokatalytischen Effekt." *Archiv für Protistekunde* 92 (1939): 453–84.

MacArthur, Robert H. "Fluctuations of Animal Populations, and a Measure of Community Stability." *Ecology* 36 (1955): 533–36.

McClung, C. E. "Vernon Lyman Kellogg, 1867–1937." *Biographical Memoirs of the National Academy of Sciences* 20 (1939): 245–57.

McIntosh, Robert P. *The Background of Ecology: Concept and Theory*. Cambridge: Cambridge University Press, 1985.

Maienschein, Jane. "Cell Lineage, Ancestral Reminiscence, and the Biogenetic Law." *Journal of the History of Biology* 11 (1978): 129–58.

———. "Introduction." In *Defining Biology. Lectures from the 1890s*, edited by Jane Maienschein, 3–50. Cambridge, Mass.: Harvard University Press, 1986.

———. "Physiology, Biology, and the Advent of Physiological Morphology." In *Physiology in the American Context, 1850–1940,* ed. Gerald L. Geison, 177–93. Bethesda, Md.: American Physiological Society, 1987.

———. "Whitman at Chicago: Establishing a Chicago Style of Biology?" In *The American Development of Biology,* edited by Ronald Rainger, Keith R. Benson, and Jane Maienschein, 151–82. Philadelphia: University of Pennsylvania Press, 1988.

Malinowski, Bronislaw. *Sex and Repression in Savage Society.* New York: Harcourt, Brace & Co., 1927.

Manning, Kenneth R. *Black Apollo of Science: The Life of Ernest Everett Just.* New York: Oxford University Press, 1983.

Marcell, David W. *Progress and Pragmatism: James, Dewey, Beard and the American Idea of Progress.* Westport, Conn.: Greenwood Press, 1974.

Marchand, C. Roland. *The American Peace Movement and Social Reform, 1898–1918.* Princeton, N.J.: Princeton University Press, 1972.

Marsden, George M. *Fundamentalism and American Culture: The Shaping of Twentieth-Century Evangelicism: 1870–1925.* New York: Oxford University Press, 1980.

Masure, Ralph H., and W. C. Allee. "Flock Organization of the Shell Parrakeet *Melopsittacus Undulatus* Shaw." *Ecology* 15 (1934): 388–98.

———. "The Social Order in Flocks of the Common Chicken and Pigeon." *Auk* 51 (1934): 306–27.

Matthews, Fred H. *Quest for an American Sociology: Robert Park and the Chicago School.* Montreal: McGill-Queen's University Press, 1977.

May, Robert M. *Stability and Complexity in Model Ecosystems.* Princeton, N.J.: Princeton University Press, 1973.

Mayr, Ernst. *The Growth of Biological Thought: Diversity, Evolution and Inheritance.* Cambridge, Mass.: Belknap Press, 1982.

———. "Speciation Phenomena in Birds." *American Naturalist* 74 (1940): 252–53.

———. "Where Are We?" In *Evolution and the Diversity of Life,* 307–28. Cambridge, Mass.: Harvard University Press, 1976.

Mayr, Ernst, and William Provine, eds. *The Evolutionary Synthesis* Cambridge, Mass.: Harvard University Press, 1980.

Mehler, Barry. "A History of the American Eugenics Society, 1921–1940." Ph.D. diss., University of Illinois, 1988.

Menge, Edward J. "Darwinism, Militarism, Socialism, and Bolschevism in the Universities." *Education* 41 (1920): 73–85.

Metcalf, Maynard M. "Darwinism and Nations." *Anatomical Record,* suppl. 14 (1918): 1–20.

Miller, E. Morton. "Caste Differentiation in the Lower Termites." In *Biology of Termites,* edited by Kumar Krishna and Frances M. Weesner, 283–310. New York: Academic Press, 1969.

———. "The Problem of Castes and Caste Differentiation in *Prorhinotermes simplex* (Hagen)." *Bulletin of the University of Miami* 15 (1942): 3–27.

Miller, Gerritt S., Jr. "Some Elements of Sexual Behavior in Primates and Their Possible Influence on the Beginnings of Human Social Development." *Journal of Mammology* 9 (1928): 273–93.

Mitchell, P. Chalmers. *Evolution and the War.* London: John Murray, 1915.

Mitman, Gregg. "Evolution as Gospel: William Patten, the Language of Democracy, and the Great War." *Isis* 81 (1990): 446–63.

Mitman, Gregg, and Richard W. Burkhardt, Jr. "Struggling for Identity: The Study of Animal Behavior in America, 1920–1945." In *The Expansion of American Biology,* edited by Keith R. Benson, Ronald Rainger, and Jane Maienschein, 164–94. New Brunswick, N.J.: Rutgers University Press, 1991.

Mitman, Gregg, and Anne Fausto-Sterling. "Whatever Happened to *Planaria?* C. M. Child and the Physiology of Inheritance." In *The Right Tools for the Job: Instruments, Materials, Techniques and Work Organization in Twentieth Century Life Sciences,* edited by Adele E. Clarke and Joan Fujimura. Princeton, N.J.: Princeton University Press, 1992.

Montagu, M. F. Ashley. "The Nature of War and the Myth of Nature." *Scientific Monthly* 54 (1942): 342–53.

———. The Socio-biology of Man." *Scientific Monthly* 50 (1940): 483–90.

———, ed., *Man and Aggression.* New York: Oxford University Press, 1968.

Moore, James R. *The Post-Darwinian Controversies: A Study of the Protestant Struggle to Come to Terms with Darwin in Great Britain and America, 1870–1900.* Cambridge: Cambridge University Press, 1979.

———. "Socializing Darwinism: Historiography and the Fortunes of a Phrase." In *Science as Politics,* edited by Les Levidow, 38–80. London: Free Association Books, 1986.

Morse, Edward S. "Charles Otis Whitman." *Biographical Memoirs of the National Academy of Sciences* 7 (1913): 269–88.

Newman, Horatio H. "History of the Department of Zoology in the University of Chicago." *Bios* 19 (1948): 215–39.

Newman, Horatio H., Frank N. Freeman, and Karl J. Holzinger. *Twins: A Study of Heredity and Environment.* Chicago: University of Chicago Press, 1937.

Nice, Margaret M. *Research is a Passion with Me: The Autobiography of Margaret Morse Nice.* Toronto: Consolidated Amethyst Communications, 1979.

———. "The Role of Territory in Bird Life." *American Midland Naturalist* 26 (1941): 441–87.

Nicholson, A. J. "The Balance of Animal Populations." *Journal of Animal Ecology* 2 (1933): 132–78.

Noble, G. K. *The Biology of the Amphibia.* 2d. ed. Dover, 1954.

———. "The Experimental Animal from the Naturalist's Point of View." *American Naturalist* 73 (1939): 113–26.

———. "The Role of Dominance in the Social Life of Birds." *Auk* 56 (1939): 263–73.

Noble, G. K., and Brian Curtis. "The Social Behavior of the Jewel Fish, *Hemichromis Bimaculatus* Gill." *Bulletin of the American Museum of Natural History* 76 (1939): 1–46.

Noble, G. K., and B. Greenberg. "Induction of Female Behavior in the Male *Anola carlinensis* with Testosterone Proprionate." *Proceedings of the Society for Experimental Biology and Medicine* 47 (1941): 32–37.

Noble, G. K., K. F. Kumpf, and V. N. Billings. "The Induction of Brooding Behavior in the Jewel Fish." *Endocrinology* 23 (1938): 353–59.

Noble, G. K., and M. Wurm. "The Effect of Testosterone Proprionate on the Black-Crowned Night Heron." *Endocrinology* 26 (1940): 837–50.

Noble, G. K., M. Wurm, and A. Schmidt. "Social Behavior of the Black-Crowned Night Heron." *Auk* 55 (1938): 7–40.

Noble, G. K., and A. Zitrin. "Induction of Mating Behavior in Male and Female Chicks Following Injections of Sex Hormones." *Endocrinology* 30 (1942): 327–34.

Novikoff, Alex B. "The Concept of Integrative Levels and Biology." *Science* 101 (1945): 209–15.

———. "Continuity and Discontinuity in Evolution." *Science* 102 (1945): 405–6.

Numbers, Ronald L. "Creationism in Twentieth-Century America." *Science* 218 (1982): 538–44.

Odum, Eugene P. "Annual Cycle of the Black-Capped Chickadee. I." *Auk* 58 (1941): 314–33.

———. *Fundamentals of Ecology.* Philadelphia: W. B. Saunders, 1959.

Oesting, R. B., and W. C. Allee. "Further Analysis of the Protective Value of Biologically Conditioned Fresh Water for the Marine Turbellarian, Procerodes Wheatlandi. IV. The Effect of Calcium." *Biological Bulletin* 68 (1935): 314–26.

Osburn, Raymond C. "Some Common Misconceptions of Evolution." *Ohio Journal of Science* 22 (1922): 173–92.

Oudshoorn, Nelly. "Endocrinologists and the Conceptualization of Sex, 1920–1940." *Journal of the History of Biology* 23 (1990): 163–86.

———. "On the Making of Sex Hormones: Research Materials and the Production of Knowledge." *Social Studies of Science* 20 (1990): 5–33.

Palmer, Ralph S. "Resolution of Respect, Dr. Charles C. Adams (1873–1955)." *Bulletin of the Ecological Society of America* 37 (1956): 103–5.

Park, Robert. *Human Communities: The City and Human Ecology,* edited by Everett C. Hughes et al. Glencoe, Ill.: Free Press, 1952.

———. "Human Nature and Collective Behavior." *American Journal of Sociology* 32 (1927): 733–41.

Park, Thomas. "Alfred Edwards Emerson, Eminent Ecologist—1967." *Bulletin of the Ecological Society of America* 48 (1967): 104–7.

———. "Beetles, Competition, and Populations." *Science* 138 (1962): 1369–75.

———. "Experimental Studies of Interspecies Competition I. Competition between Populations of the Flour Beetles, *Tribolium confusum* Duval and

Tribolium castaneum Herbst." *Ecological Monographs* 18 (1948): 265–308.

———. "Integration in Infra-social Insect Populations." In *Levels of Integration in Biological and Social Systems,* edited by Robert Redfield, 121–38. Lancaster, Penn.: Jaques Cattell Press, 1942.

———. "The Laboratory Population as a Test of a Comprehensive Ecological System." *Quarterly Review of Biology* 16 (1941): 274–93, 440–61.

———. "Some Observations on the History and Scope of Population Ecology." *Ecological Monographs* 16 (1946): 315–20.

———. "Studies in Population Physiology. II. Factors Regulating Initial Growth of Tribolium Confusum Populations." *Journal of Experimental Zoology* 65 (1933): 17–42.

———. "Studies in Population Physiology: The Relation of Numbers to Initial Population Growth in the Flour Beetle *Tribolium Confusum* Duval." *Ecology* 13 (1932): 172–81.

Patten, William. *Evolution.* Hanover, N.H.: Dartmouth Press, 1924.

———. *The Grand Strategy of Evolution.* Boston: Richard G. Badger, 1920.

———. *Growth: Introduction to the Study of Evolution.* Hanover, N.H.: Dartmouth Press, 1923.

———. "The Message of the Biologist." *Science* 51 (1920): 93–102.

Pattison, William D. "Goode's Proposal of 1902: An Interpretation." *Professional Geographer* 30 (1978): 3–8.

———. "Rollin Salisbury and the Establishment of Geography at the University of Chicago." In *The Origins of Academic Geography in the United States,* edited by Brian W. Blouet, 151–64. Hamden, Conn.: Archon Books, 1981.

Paul, Diane B. "The Rockefeller Foundation and the Origins of Behavior Genetics." In *The Expansion of American Biology,* edited by Keith R. Benson, Ronald Rainger, and Jane Maienschein, 262–83. New Brunswick, N.J.: Rutgers University Press, 1991.

Pauly, Philip J. "The Appearance of Academic Biology in Late Nineteenth-Century America." *Journal of the History of Biology* 17 (1984): 369–97.

———. *Controlling Life: Jaques Loeb and the Engineering Ideal in Biology.* New York: Oxford University Press, 1987.

———. "General Physiology and the Discipline of Physiology," in *Physiology in the American Context, 1850–1940,* ed. Gerald L. Geison, 195–207. Bethesda, Md.: American Physiological Society, 1987.

———. "The Loeb-Jennings Debate and the Science of Animal Behavior." *Journal of the History of the Behavioral Sciences* 17 (1981): 504–15.

———. "Summer Resort and Scientific Discipline: Woods Hole and the Structure of American Biology, 1882–1925." In *The American Development of Biology,* edited by Ronald Rainger, Keith R. Benson, and Jane Maienschein, 121–50. Philadelphia: University of Pennsylvania Press, 1988.

Pearl, Raymond. "Biology and War." *Journal of the Washington Academy of Sciences* 8 (1918): 341–60.

———. *The Biology of Population Growth*. New York: Alfred A. Knopf, 1925.

———. "The Evolution of Sociality." *Ecology* 20 (1939): 305–10.

———. "On Biological Principles Affecting Populations: Human and Other." *American Naturalist* 71 (1937): 50–68.

Pearl, Raymond, John Rice Miner, and Silvia L. Parker. "Experimental Studies on the Duration of Life. XI. Density of Population and Life Duration in Drosophila." *American Naturalist* 61 (1927): 289–318.

Pearse, A. S. *Animal Ecology*. New York: McGraw-Hill Book Co., 1926.

Pells, Richard H. *The Liberal Mind in the Conservative Age: American Intellectuals in the 1940s and 1950s*. New York: Harper & Row, 1985.

———. *Radical Visions and American Dreams*. New York: Harper & Row, 1973; Middletown, Conn.: Wesleyan University Press, 1984.

Petersen, Walburga A. "The Relation of Density of Population to Rate of Reproduction in Paramecium Caudatum." *Physiological Zoology* 2 (1929): 221–54.

Phipps, C. F. "An Experimental Study of the Behavior of Amphipods with Respect to Light Intensity, Direction of Rays and Metabolism." *Biological Bulletin* 28 (1915): 210–23.

Pickens, Donald. *Eugenics and the Progressives*. Nashville, Tenn.: Vanderbilt University Press, 1968.

Pigors, Paul. *Leadership or Domination*. New York: Houghton Mifflin, 1935.

Price, David E. "Community and Control: Critical Democratic Theory in the Progressive Period." *American Political Science Review* 4 (1974): 1662–78.

Provine, William B. *The Origins of Theoretical Population Genetics*. Chicago: University of Chicago Press, 1971.

———. "The Role of Mathematical Population Genetics in the Evolutionary Synthesis of the 1930s and 1940s." *Studies in the History of Biology* 2 (1978): 167–92.

———. *Sewall Wright and Evolutionary Biology*. Chicago: University of Chicago Press, 1986.

Purcell, Edward A., Jr. *The Crisis of Democratic Theory: Scientific Naturalism and the Problem of Value*. Lexington: University Press of Kentucky, 1973.

Quandt, Jean B. *From the Small Town to the Great Community: The Social Thought of Progressive Intellectuals*. New Brunswick, N.J.: Rutgers University Press, 1970.

Rainger, Ronald. "Vertebrate Paleontology as Biology: Henry Fairfield Osborn and the American Museum of Natural History." In *The American Development of Biology*, edited by Ronald Rainger, Keith R. Benson, and Jane Maienschein, 219–56. Philadelphia: University of Pennsylvania Press, 1988.

Redfield, Robert, ed. *Levels of Integration in Biological and Social Systems*. Biological Symposia, vol. 8. Lancaster, Pa.: Jaques Cattell Press, 1942.

Reuter, E. B. "The Relation of Biology and Sociology." *American Journal of Sociology* 32 (1927): 705–18.

Richards, Robert J. *Darwin and the Emergence of Evolutionary Theories of Mind and Behavior.* Chicago: University of Chicago Press, 1987.

Riesman, David. *Individualism Reconsidered and Other Essays.* Glencoe, Ill.: Free Press, 1954.

Ritchey, Frances. "Dominance-Subordination and Territorial Relationships in the Common Pigeon." *Physiological Zoology* 20 (1951): 167–76.

Ritter, William E. "Biology's Contribution to a System of Morals That Would Be Adequate for Modern Civilization." *Bulletin of the Scripps Institute of Biological Research* 2 (1917): 3–8.

———. "A Business Man's Appraisement of Biology." *Science* 44 (1916): 819–22.

———. *The Unity of the Organism.* 2 vols. Boston: R. G. Badger, 1919.

———. *War, Science and Civilization.* Boston: Sherman, French & Co., 1915.

Roback, E. "Theodore C. Schneirla, 1902–1968." *American Journal of Psychology* 94 (1981): 355–57.

Robertson, T. Brailsford. "CLXII. Allelocatalytic Effect in Cultures of *Colpidium* in Hay-Infusion and in Synthetic Media." *Biochemical Journal* 18 (1924): 1240–47.

———. *The Chemical Basis of Growth and Senescence.* Philadelphia: J. B. Lippincott Co., 1923.

———. "II. The Influence of Mutual Contiguity upon Reproductive Rate and the Part Played Therein By the 'X-Substance' in Bacterised Infusions Which Stimulates the Multiplication of Infusoria." *Biochemical Journal* 15 (1921): 612–19.

———. "The Influence of Washing upon the Multiplication of Isolated Infusoria and upon Allelocatalytic Effect in Cultures Initially Containing Two Infusoria." *Australian Journal of Experimental Biology and Medicine* 1 (1924): 151–73.

———. "I. The Multiplication of Isolated Infusoria." *Biochemical Journal* 15 (1921): 595–611.

———. "On Some Conditions Affecting the Viability of Cultures of Infusoria and the Occurrence of Allelocatalysis Therein." *Australian Journal of Experimental Biology and Medicine* 4 (1927): 1–24.

———. "Reproduction in Cell-Communities." *Journal of Physiology* 56 (1922): 404–12.

Rodgers, Andrew Denny, III. *John Merle Coulter: Missionary in Science.* Princeton, N.J.: Princeton University Press, 1944.

Rodgers, Daniel. "In Search of Progressivism." *Reviews in American History* 10 (1982): 113–32.

Rossiter, Margaret W. *Women Scientists in America: Struggles and Strategies to 1940.* Baltimore: Johns Hopkins University Press, 1982.

Rucker, Darnell. *The Chicago Pragmatists.* Minneapolis: University of Minnesota Press, 1969.

Russett, Cynthia Eagle. *The Concept of Equilibrium in American Social Thought*. New Haven, Conn.: Yale University Press, 1966.

Ryan, W. Carson. *Studies in Early Graduate Education: The Johns Hopkins, Clark University, The University of Chicago*, 1939. Reprint. New York: Arno Press and the *New York Times*, 1971.

Salisbury, Rollin D. "Geology in Education." *Science* 47 (1918): 325–35.

Salisbury, Rollin D., and William C. Alden. "The Geography of Chicago and Its Environs." *Bulletin of the Geographic Society of Chicago* 1 (1899): 1–64.

Salisbury, Rollin D., Harlan H. Barrows, and Walter S. Tower, *The Elements of Geography*. New York: Henry Holt & Co., 1912.

Salt, George. *Report on Sugar-Cane Borers at Soledad, Cuba*. Contributions From the Harvard Institute for Tropical Biology and Medicine, III. Cambridge, Mass.: Harvard University Press, 1926.

Sapp, Jan. *Beyond the Gene: Cytoplasmic Inheritance and the Struggle for Authority in Genetics*. New York: Oxford University Press, 1987.

Scherba, Gerald M. "Microclimatic Modification as a Homeostatic Mechanism in Ant Mounds." Ph.D. diss., University of Chicago, 1955.

Schjelderup-Ebbe, T. "Beitrage zur Sozial-psychologie des Haushuhns." *Zeitschrift für Psychologie* 88 (1922): 222–25.

———. "Social Behavior of Birds." In *A Handbook of Social Psychology*, ed. Carl Murchison, 947–72. Worcester, Mass.: Clark University Press, 1935.

Schmidt, Karl Patterson. "Warder Clyde Allee, 1885–1955." *Biographical Memoirs of the National Academy of Sciences* 30 (1957): 3–40.

Schmidt, Robert S. "The Evolution of Nest-Building Behavior in *Apicotermes* (Isoptera)." *Evolution* 9 (1955): 157–81.

Schneirla, Theodore C. "Problems in the Biopsychology of Social Organization." *Journal of Abnormal Social Psychology* 41 (1946): 385–402.

Scott, John P. *Aggression*. Chicago: University of Chicago Press, 1958.

———. "Dominance and the Frustration-Aggression Hypothesis." *Physiological Zoology* 21 (1948): 31–39.

———. "The Embryology of the Guinea Pig. III. Development of the Polydactylous Monster. A Case of Growth Accelerated at a Particular Period by a Semi-dominant Lethal Gene." *Journal of Experimental Zoology* 77 (1937): 123–57.

———. "An Experimental Test of the Theory that Social Behavior Determines Social Organization." *Science* 99 (1944): 42–43.

———. "Genetic Differences in the Social Behavior of Inbred Strains of Mice." *Journal of Heredity* 35 (1942): 11–15.

———. "Investigative Behavior: Toward a Science of Sociality." In *Studying Animal Behavior: Autobiographies of the Founders*, edited by Donald A. Dewsbury, 389–430. Chicago: University of Chicago Press, 1985.

———. "The Organization of Comparative Psychology," *Annals of the New York Academy of Science* 223 (1973): 7–40.

———. "Social Behavior, Organization and Leadership in a Small Flock of

Domestic Sheep." *Comparative Psychology Monographs* 18 (1945): 1–29.

Scott, John P., and John L. Fuller. "Research on Genetics and Social Behavior at the Roscoe B. Jackson Memorial Laboratory, 1946–1951—A Progress Report." *Journal of Heredity* 42 (1951): 191–96.

Scott, J. W. "Mating Behavior of the Sage Grouse." *Auk* 59 (1942): 477–98.

Sears, Paul B. "Charles C. Adams, Ecologist." *Science* 123 (1956): 974.

Shaw, Gretchen. "The Effect of Biologically Conditioned Water upon Rate of Growth in Fishes and Amphibia." *Ecology* 13 (1932): 263–278.

Shelford, Victor E. *Animal Communities in Temperate America*. Chicago: University of Chicago Press, 1913.

———. "A Comparison of the Responses of Sessile and Motile Plants and Animals." *American Naturalist* 48 (1914): 641–74.

———. "Ecological Succession. I. Stream Fishes and the Method of Physiographic Analysis." *Biological Bulletin* 11 (1911): 9–35.

———. "Ecological Succession. II. Pond Fishes." *Biological Bulletin* 11 (1911): 127–51.

———. "Ecological Succession. III. A Reconnaissance of Its Causes in Ponds with Particular Reference to Fish." *Biological Bulletin* 12 (1911): 1–38.

———. "Ecological Succession. IV. Vegetation and the Control of Land Communities." *Biological Bulletin* 13 (1912): 59–99.

———. "Ecological Succession. V. Aspects of Physiological Classification." *Biological Bulletin* 23 (1912): 331–70.

———. "An Experimental Study of the Behavior Agreement among the Animals of an Animal Community." *Biological Bulletin* 26 (1914): 294–315.

———. "Life Histories and Larval Habits of the Tiger Beetles (*Cicindelida*)." *Journal of the Linnaean Society of London* 30 (1908): 157–84.

———. "Physiological Animal Geography." *Journal of Morphology* 22 (1911): 551–618.

———. "Preliminary Note on the Distribution of the Tiger Beetles (Cicendela) and Its Relation to Plant Succession." *Biological Bulletin* 14 (1908): 9–14.

———. "Principles and Problems of Ecology as Illustrated by Animals." *Journal of Ecology* 3 (1915): 1–23.

Shelford, Victor E., and W. C. Allee. "An Index of Fish Environments." *Science* 36 (1912): 76–77.

———. "Rapid Modification of Behavior of Fishes by Contact with Modified Water." *Animal Behavior* 4 (1914): 1–30.

———. "The Reaction of Fishes to Gradients and Dissolved Atmospheric Gases." *Journal of Experimental Zoology* 14 (1913): 207–63.

Shoemaker, Hurst H. "Social Hierarchy in Flocks of the Canary." *Auk* 56 (1939): 381–406.

Simberloff, Daniel. "A Succession of Paradigms in Ecology: Essentialism to Materialism and Probabilism." In *Conceptual Issues in Ecology*, edited by Esa Saarinen, 63–100. Dordrecht: D. Reidel Publishing Co., 1980.

Simpson, George Gaylord. *The Meaning of Evolution*. New Haven, Conn.: Yale University Press, 1949.

———. "The Role of the Individual in Evolution." *Journal of the Washington Academy of Sciences* 31 (1941): 1–20.

Sinnott, Edmund W. "The Biological Basis of Democracy." *Yale Review* 35 (1945–1946): 61–73.

Smith, Harry S. "The Role of Biotic Factors in the Determination of Population Densities." *Journal of Economic Entomology* 28 (1935): 873–98.

Smith, John Maynard. "Group Selection." *Quarterly Review of Biology* 51 (1976): 277–83.

Smith, T. V., and Leonard White, eds. *Chicago: An Experiment in Social Science Research*. Chicago: University of Chicago Press, 1929.

Snyder, Thomas Elliott. "The Biology of the Termite Castes." *Quarterly Review of Biology* 1 (1926): 522–52.

Sober, Elliott R. *The Nature of Selection: Evolutionary Theory in Philosophical Focus*. Cambridge, Mass.: MIT Press, 1984.

Sonneborn, Tracy. "Herbert Spencer Jennings." *Biographical Memoirs of the National Academy of Sciences* 47 (1975): 143–223.

Spencer, Herbert. *Principles of Sociology,* vols. 1–2. Authorized edition. New York: Appleton, 1897.

Spoehr, Luther William. "Progress' Pilgrim: David Starr Jordan and the Circle of Reform, 1891–1931." Ph.D. diss., Stanford University, 1975.

Stepan, Nancy Leys. "'Nature's Pruning Hook': War, Race and Evolution, 1914–1918." In *The Political Culture of Modern Britain: Studies in Memory of Stephen Koss,* edited by J. M. W. Bean, 129–48. London: 1987.

———. "Race and Gender: The Role of Analogy in Science." *Isis* 77 (1986): 261–77.

Stewart, Jeannie C., and J. P. Scott. "Lack of Correlation between Dominance Relationships in a Herd of Goats." *Journal of Comparative and Physiological Psychology* 40 (1947): 255–64.

Storr, Richard J. *Harper's University: The Beginnings*. Chicago: University of Chicago Press, 1966.

Sweet, Helen. "A Micropopulation of *Euglena Gracilis* Klebs in Sterile Autotrophic Media and in Bacterial Suspensions." *Physiological Zoology* 12 (1939): 173–208.

Talbot, Mary. "Distribution of Ant Species in the Chicago Region with Reference to Ecological Factors and Physiological Toleration." Ph.D. diss., University of Chicago, 1934.

Tarde, Gabriel. "Inter-psychology." *International Quarterly* 7 (1903): 59–84.

Taylor, Norman. "Some Modern Trends in Ecology." *Torreya* 12 (1912): 110–17.

Taylor, Peter J. "Technocratic Optimism, H. T. Odum, and the Partial Transformation of Ecological Metaphor after World War II." *Journal of the History of Biology* 21 (1988): 213–44.

Thomas, Norman. *The Conscientious Objector in America.* New York: B. W. Huebsch, 1923.

Thompson, C. B. "Origin of the Castes of the Common Termite, *Leucotermes Flavipes* Kollar." *Journal of Morphology* 30 (1917): 83–153.

———. "The Development of the Castes of Nine Genera and Thirteen Species of Termites." *Biological Bulletin* 36 (1919): 379–98.

Thompson, William R. "Biological Control and the Theories of the Interactions of Populations." *Parasitology* 31 (1939): 299–388.

Tinbergen, Nikolaas. *Social Behavior in Animals with Special Reference to Vertebrates.* New York: John Wiley & Sons, 1953.

Tobach, Ethel, and Lester R. Aronson, "Biographical Note." In *Development and Evolution of Behavior,* xi–xviii. San Francisco: W. H. Freeman & Co., 1970.

Tobey, Ronald C. *Saving the Prairies: The Life Cycle of the Founding School of American Plant Ecology, 1895–1955.* Berkeley and Los Angeles: University of California Press, 1981.

Todes, Daniel P. *Darwin without Malthus: The Struggle for Existence in Russian Evolutionary Thought.* New York: Oxford University Press, 1989.

Tower, Walter S. "Scientific Geography: The Relation of Its Content to Its Subdivisions." *Bulletin of the American Geographical Society.* 42 (1910): 801–25.

Trivers, Robert L. "The Evolution of Reciprocal Altruism." *Quarterly Review of Biology* 46 (1971): 35–57.

Turner, Charles Henry. "The Homing of Ants: An Experimental Study of Ant Behavior." Ph.D. diss., University of Chicago, 1907.

Unesco. *The Race Concept: Results of an Inquiry.* Paris: Unesco, 1952.

Uvarov, B. P. "Insects and Climate." *Transactions of the Entomological Society* 79 (1931): 1–247.

Violas, Paul C. "Progressive Social Philosophy: Charles Horton Cooley and Edward Alsworth Ross." In *Roots of Crisis: American Education in the Twentieth Century,* edited by Clarence J. Karier, Paul C. Violas, and Joel Spring, 40–65. Chicago: Rand McNally College Publishing Co., 1973.

Waddington, C. H. *Science and Ethics.* London: George Allen & Unwin Ltd., 1942.

Wade, Michael J. "A Critical Review of the Models of Group Selection." *Quarterly Review of Biology* 53 (1978): 101–14.

Weinstein, A. "A Note on W. L. Tower's *Leptinotarsa* Work." In *The Evolutionary Synthesis,* edited by E. Mayr and W. Provine, 352–353. Cambridge: Harvard Univ. Press, 1980.

Weldon, Stephen. "Integrating the Ant: T. C. Schneirla's Comparative Psychology." Master's paper, Department of History of Science, University of Wisconsin—Madison, 1990.

Wells, Morris M. "The Reactions and Resistance of Fishes in Their Natural Environment to Salts." *Journal of Experimental Zoology* 19 (1915): 243–83.

Wheeler, William Morton. "Animal Societies." *Scientific Monthly* 39 (1934): 289–301.

———. *Emergent Evolution and the Development of Societies.* New York: W. W. Norton & Co., 1928.

———. *Mosaics and Other Anomalies among Ants.* Cambridge, Mass.: Harvard University Press, 1937.

———. *The Social Insects.* New York: Harcourt, Brace & Co., 1928.

———. *Social Life among the Insects.* New York: Harcourt, Brace & Co., 1923.

———. "Societal Evolution." In *Human Biology and Racial Welfare,* edited by E. V. Cowdry, New York: Paul Hoeber, Inc., 1930.

Whitman, Charles Otis. "Animal Behavior." *Biological Lectures Delivered at the Marine Biological Laboratory of Wood's Holl,* 285–338. 1899.

———. "Biological Instruction in the Universities." *American Naturalist* 21 (1887): 507–19.

———. "The Inadequacy of the Cell-Theory of Development." *Journal of Morphology* 8 (1893): 639–58.

———. "Some of the Functions and Features of a Biological Station." *Biological Lectures Delivered at the Marine Biological Laboratory of Wood's Holl,* 240–41. 1899.

———. "Specialization and Organization." *Biological Lectures Delivered at the Marine Biological Laboratory of Wood's Holl,* 1–26. 1891.

Williams, George C. *Adaptation and Natural Selection: A Critique of Some Current Evolutionary Thought.* Princeton, N.J.: Princeton University Press, 1966.

Willier, Benjamin H. "Frank Rattray Lillie, 1870–1947." *Biographical Memoirs of the National Academy of Sciences* 30 (1957): 179–236.

Wilson, David Sloan. "The Group Selection Controversy: History and Status." *Annual Review of Ecology and Systematics* 14 (1983): 159–87.

Wilson, Edward O. "Group Selection and Its Significance for Ecology." *BioScience* 23 (1973): 631–38.

———. *On Human Nature.* Cambridge, Mass.: Harvard University Press, 1978.

———. *Sociobiology: The New Synthesis.* Cambridge, Mass.: Harvard University Press, 1975.

Wilson, Edward O. and Charles D. Michener. "Alfred Edwards Emerson, December 31, 1896–October 3, 1976." *Biographical Memoirs of the National Academy of Sciences* 53 (1982): 159–77.

Winslow, Amy. "Marjorie Hill Allee, June 2, 1890–April 30, 1945." *Horn Book Magazine* 22 (1946): 183–95.

Worster, Donald. *Nature's Economy: The Roots of Ecology.* Garden City, N.J.: Anchor Press, 1979.

Wright, Sewall. "Evolution in Mendelian Populations." *Genetics* 16 (1931): 97–159.

———. "Genetics of Abnormal Growth in the Guinea Pig." *Cold Spring Harbor Symposia on Quantitative Biology,* 2 (1934): 137–47.

———. "Review of *The Origin and Development of the Nervous System,* by C. M. Child." *Journal of Heredity* 12 (1921): 72–75.

———. "The Role of Mutation, Inbreeding, and Crossbreeding and Selection in Evolution." *Proceedings of the Sixth International Congress of Genetics* 1 (1932): 356–66.

Wyllie, Irvin G. "Social Darwinism and the American Businessman." *Proceedings of the American Philosophical Society* 103 (1959): 629–35.

Wynne-Edwards, V. C. *Animal Dispersion in Relation to Social Behavior.* Edinburgh: Oliver & Boyd, 1962.

Yerkes, Robert M., ed. *The New World of Science: Its Development during the War.* New York: Century Press, 1920.

Young, Robert M. *Darwin's Metaphor: Nature's Place in Victorian Culture.* Cambridge: Cambridge Univ. Press, 1985.

———. "The Naturalization of Value Systems in the Human Sciences." In *Science and Belief: Darwin to Einstein.* Block VI: *Problems in the Biological and Human Sciences,* 63–110. Milton Keynes: Open University Press, 1981.

Zuckerman, Solly. *The Social Life of Monkeys and Apes.* London: Routledge & Kegan Paul, 1932.

Index